MRP

INTEGRATING MATERIAL REQUIREMENTS PLANNING AND MODERN BUSINESS

The IRWIN PROFESSIONAL PUBLISHING/APICS Series in Production Management

Supported by the American Production and Inventory Control Society

MRP

INTEGRATING MATERIAL REQUIREMENTS PLANNING AND MODERN BUSINESS

Terry Lunn, CFPIM
with
Susan A. Neff, CPIM

Foreword by George W. Plossl, CFPIM

IRWIN
Professional Publishing
Burr Ridge, Illinois
New York, New York

Definitions from the APICS Dictionary are reprinted with permission, The American Production and Inventory Control Society, APICS Dictionary, Sixth Edition, 1987.

Sponsoring editor: Jean Marie Geracie
Project editor: Waivah Clement
Production manager: Mary Jo Parke
Designer: Jeanne M. Rivera
Art manager: Kim Meriwether
Art: Graphics Plus
Compositor: TCSystems, Inc.
Typeface: 11/13 Palatino
Printer: Book Press, Inc.

Library of Congress Cataloging-in-Publication Data

Lunn, Terry.
MRP : integrating material requirements planning and modern business / Terry Lunn with Susan A. Neff.
p. cm.—(Business One Irwin/APICS series in production management)
ISBN 1-55623-656-5
1. Inventory control—Data processing. 2. Production control—Data processing. I. Neff, Susan A. II. Title. III. Series.
TS160.L86 1992
658.7'87—dc20 92–1389

Printed in the United States of America
2 3 4 5 6 7 8 9 0 BP 9 8 7 6 5 4 3

For our mothers, Lenore and Liz

FOREWORD

When people do not fully understand how their computer-based systems work, they take one of two courses: follow them blindly or ignore them completely. Both are guarantees of failure to gain the full potential benefits.

Planners working with MRP programs need to understand more than the mechanics of the system; they must also know what latitude they have to override it and change the actions suggested. Many others, however, including cost accountants, design and process engineers, production workers, purchasing people, and clerical workers in internal sales activities, need to understand only how the MRP programs work and what information they can provide them.

The literature on MRP—material requirements planning—falls into two classes: in-depth concentration on the technique and its manifold components or superficial descriptions of it and its role in manufacturing planning and control. The former turns many people off; like the boy who asked the librarian for a book on zebras and, when offered a thick tome, said, "I don't want to know that much about them." The latter does not provide the basic working knowledge many nondirect users need.

This book bridges the gap, enabling support staff people to quickly and easily gain the basic knowledge they need. It also provides a fine primer for beginning the education of planners. It covers the mechanics of MRP in clear, concise language and shows its application to many types of production.

Throughout the text, Focus boxes direct attention to specific actions that enhance or endanger the successful use of the techniques. At the end of each chapter, an Applications section illustrates how the techniques are being used in a variety of types of businesses, thus broadening readers' understanding and alerting them to potential problems.

This book is useful both as a text for educational programs and as a reference for practitioners in industry using MRP programs or interrelating with them in support activities. It has useful information for those implementing systems. The text is well written, figures are excellent, and appendixes provide details on supplemental reading, hands-on MRP exercises, descriptions of lot-sizing techniques, and a discussion of the closed-loop system of which MRP is a core element.

The publication of this book fills an empty niche and meets a long-standing need. It is a valuable addition to the literature of the field of manufacturing planning and control.

George W. Plossl

PREFACE

Material requirements planning is the scheduling logic used to manage the flow of material in manufacturing operations. Every manufacturing company in the country is reevaluating its ability to compete, striving to find ways to educate and motivate its employees toward excellence. The efficient flow of material is a major objective of this drive toward excellence. We concluded that there is a vital need for a basic and approachable book on the subject of material requirements planning (MRP).

We are going to cut through the academic and technical jargon that has separated end users from the system, clearly tracing the logical system of MRP in easy terms and examples. Although material requirements planning is commonly computer-implemented, its basic principles can conceivably be applied manually in small business organizations. The ability to understand its essential premises is a prerequisite to the successful implementation of any manufacturing control system.

The essential components of a material requirements planning system are presented in simple sequential steps. The first nine chapters of the book deal with the inputs into the process and how this raw data is converted into meaningful management information. As far as possible, we have tried to use examples from the work life of a small plant that produces pens and pencils—products easily visualized by the reader. We will examine how employees are struggling to meet production schedules, satisfy customer requirements, and control operating costs—laudable goals that sometimes appear to be mutually exclusive. The reader will recognize problems and constraints that are familiar incidents in day-to-day plant operations and will learn how to use and apply the powerful tools of a material requirements planning system to solve scheduling dilemmas and control the manufacturing environment.

Standardized definitions as recognized within the manufacturing community will be used throughout this book. We have indicated in the index those terms that are under the copyright of the American Production and Inventory Control Society (APICS).

We have included an appendix of supplemental readings that may help the reader to understand the fine detail of some of the topics we discuss. No book (even this one) can provide all of the experience and perspective we need to acquire on all aspects of the manufacturing process; we urge the reader to seek out as many learning experiences as possible to stay competitive.

To the best of our knowledge, FineLine Industries is wholly imaginary, but we kind of like to think of it as being located a couple of miles outside a small town someplace in West Tennessee.

This is *your* book, and we hope it helps you to achieve your goals. If you have any suggestions on how we might improve it, or if there was anything that was particularly useful to you, we'd like to hear from you.

Terry Lunn
Susan A. Neff

ACKNOWLEDGMENTS

No book is written in a vacuum, and this one is no exception. Without the contributions of countless people, this project would never have gotten past the stage of late-night discussions at various technical seminars.

First, and perhaps most important, are all the members and friends of the American Production and Inventory Control Society. This book was written with them in mind, and in specific response to their many pleas that Terry put the information from his seminars and classes into written form.

Connie Bogue, Mark Barnes, CPIM, L. James Burlingame, CPIM, Dennis J. Dureno, CPIM, C.P.M., Robert A. Osborne, CPIM, George W. Plossl, CFPIM, Paul J. Rosa, Jr., CFPIM, and George N. Wells, CPIM, read our drafts and made countless invaluable suggestions. The strong points in the book are augmented by their thoughtful criticism, and we take sole responsibility for any weaknesses.

Jean M. Geracie, A. Drew Gierman, and Jeffrey A. Krames of Irwin Professional Publishing led us through the unfamiliar territory of authorship with encouragement, and William T. Walker, CFPIM, gave us practical advice to help us along.

Marcia A. Brown, Charles G. Mertens, Paul E. Sheehan, CPIM, and Michael J. Stack of the APICS headquarters staff provided support and encouragement. Thanks are also due to the members of the APICS education department—and in particular Rae Hurley and Kristen Shiveley—who covered Sue's desk and fielded some of her workload during the heaviest writing periods.

Lynn Hopkins provided great support, including preparation of many of the graphics. She was also very patient about Sue's invading her office and messing around interminably with the default settings on her word processor.

We also thank our families and friends, who may have been neglected during the writing of this book but were never forgotten. Special appreciation goes to Ronn for his patience during all the weekends and holidays when writing trips took precedence over everything else. And it was Ronn who reminded us that replacement parts for the competition's product also count as independent demand items.

T.L.
S.A.N.

CONTENTS

INTRODUCTION

People working together are the heart of any successful business enterprise today. Material Requirements Planning (MRP) is a scheduling technique used by many manufacturing firms as the tool through which people communicate to one another about the flow of material. Successful companies realize that it is of primary importance for people using this tool to understand both the technique and their role in its execution. There is a real need for today's businesses, across all functional lines, to understand the workings of a material requirements planning system. Once business professionals understand the simple concepts of the MRP technique and how its parts work together, they can effectively apply this tool to make their lives easier tomorrow.

WHAT IS MRP?

Material Requirements Planning is a phrase that puts the primary emphasis on the word *planning.* It is, above all else, a planning and scheduling technique. Much confusion surrounds the concept, however, because the acronym MRP can mean several different things, including material requirements planning, closed-loop MRP, and MRP II. *MRP* in its most restricted form refers to the mathematical logic used to plan material amounts and to determine the date that those materials are needed. However, anyone in manufacturing will quickly tell you that it is what we actually do with those plans and schedules that determines success. A planning system feeds an execution system—such as shop floor control or production activity control.

Success with a planning system is an iterative process. As our plans drive the execution system, conditions begin to change over time. As these conditions change, it is necessary to have a feedback loop to manage the plan. When people are unable to execute according to the plan, we must consider the options. First and foremost, have we exhausted all possible alternatives to achieve our plan? If so, then we must tell the truth to the system and change the plan accordingly. This is called *closing the loop*. We will contrast the steps necessary to complete the execution of the plan with the steps that are necessary when replanning must be performed.

MRP planning is typically used to schedule the flow of materials to and through a manufacturing process. When these plans are accurate and we use them for execution, we are then able to use the materials flow as a base for planning all the company's resources. For example, we can compute the money required for raw materials by using future purchase expectations. We can calculate capacity required, which extends to labor needs, which extends, in turn, to payroll requirements. We can examine those resources that may take some time to modify, such as equipment, space, and energy requirements, so that we can look ahead to an orderly, structured adjustment in these areas. In total, we can extrapolate our material plans into financial plans.

When our planning becomes this extensive, it is called *MRP II*, or *manufacturing resource planning*. Generally, in the discussions that follow, when we refer to an MRP system, we are referring to this MRP II concept. We create and maintain the plans with MRP planning logic, we close the loop as execution progresses, and we use these numbers to run our total business enterprise.

At all times, we must keep in mind the essential premise that MRP is our tool, not our master. We control the tool with parameters as we make our operating decisions, and it suggests the actions we should take to implement these decisions.

WHOSE MRP IS IT?

There is one key to success in any MRP system: people. How well do you and I understand our role in this complete planning system? To help us visualize this, the book is divided into five sections. The first section, Process, reviews the basic MRP function: How does the logic of MRP planning work? In the second section, Inputs, we discuss the elements necessary as inputs to the basic MRP function. Third, we will examine the Outputs and how they are used. Once we understand the inputs and outputs, we will then turn our attention to the fourth area, Managing with MRP. We will conclude with the fifth section, People, the ultimate owners and users of the system.

HOW DO WE LOOK AT MRP?

In its abstract form, MRP applies to all business enterprises. There are certain business types, however, in which it is more easily applied. Its logic is more evident, for example, in a manufacturing company producing material in discrete lots, in which we include the range of make-to-order, job shop, make-to-stock, and repetitive manufacturing.

For purposes of effective discussion, we will use this environment to describe the MRP concepts. We will use an uncomplicated product—pens and pencils—to illustrate the key concepts and their applications. Throughout our discussion, we will consistently develop our illustrations through the manufacture of pens and pencils, products to which all readers can relate. We will show that if you do this well in one plant and one operation, you can go on, using other technologies, to link several plants or operations using the same essential concepts.

When appropriate, we will apply specific concepts to other industry types: continuous flow process, distribution, and service industries. This will enable you, the reader, to stretch your mind to envision how to apply these concepts to help improve your own operation. These Applications sections do not go into exhaustive detail about *how* certain well-run companies are moving toward excellence through creative adaptation of MRP principles; this is, after all, a book on how to perform the basic operations effectively. What we have tried to do in these sections is to kindle your imagination and give you a glimpse of what can be done once you've started that journey toward excellence. When you are farther down the road, having really internalized the basics, you should be able to see the unique things that can be done within your own company.

We have not used, in our examples and discussions, many defense-oriented businesses that are required to comply with guidelines issued by the Department of Defense (DOD). For many years, there was a misapprehension that resulted in a sort of "myth" that MRP was incompatible with DOD requirements.

Quite frankly, this line of argument is simply not true. What *is* true is that to meet DOD guidelines, some special requirements and assumptions must be built into the process and the system. MRP, for example, is less applicable in prototyping situations but is extremely useful in managing the flow of material to support long-term and repetitive contractual commitments.

As you read this book, remember that thousands of people in all functional areas of every imaginable business environment are working with MRP systems. You don't need to be a computer expert to be able to understand and use this powerful tool. We're going to walk through the basic logic with you, show you some common applications, and point out some neat tricks that creative users have put into place in their companies. Perhaps one of their ideas will light a spark for you.

SECTION 1

PROCESS

CHAPTER 1

PLANNING

To paraphrase a famous axiom, planning accomplishes nothing, but to have a plan is everything. Nowhere is this truer than in a business enterprise motivated to make a profit.

Every business provides either a product or a service that fulfills a need for the customer. Our focus has to be on the customer's need and how the output of our organization supplies that need. There must be some demand for our product or service in order for us to stay in business. Our plan, then, must have as its first consideration that output.

The second consideration is the inputs required to enable us to produce the output. Here we must examine all of the resources it takes for us to operate our company.

The third thing we need to consider is the process by which we convert our inputs into output. How do we take all of our various input resources and mesh them together to fill the customer's requirement? This is what we call integrating the resources. We demonstrate this principle in Figure 1–1.

Let's look at a couple of examples of what we mean by this. We are all familiar with discrete manufacturing—for instance, an automobile manufacturer whose output is a finished vehicle. The obvious inputs are the steel, rubber, glass, and plastic components and raw materials that it takes to make a car. But if we use the term *resources* in the broader sense of inputs, it means *anything* required for the production of a product, or any element that would cause the process to fail if it were omitted. So, we should consider the people that it takes to run the factory as a resource. Carrying this concept a step farther, resources

FIGURE 1–1
System

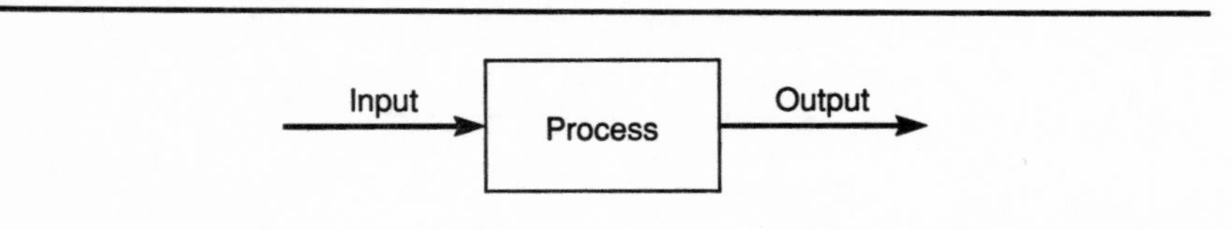

would include electrical power, money, equipment, and services such as transportation and communications.

We can easily recognize the applicability of this principle to high-volume process industries as well: steel, pharmaceuticals, petroleum distillation, and foods.

We can also apply the model to a service industry such as airfreight forwarding. Here, the output is the relocation of goods from one point to another, and the inputs are the packaging materials, airplanes, trucks, and the people who handle the packages.

One example that we intend to use throughout this text as we examine these concepts is the manufacture of pens and pencils. We will consider ourselves to be working for FineLine Industries. Our customers demand mechanical pens and pencils from us much like the ones you probably have in your pocket right now. Our inputs are steel, pieces of lead, erasers, and all the other materials that it takes to make and package either a pen or a pencil. And our inputs also include the people in the plant, equipment, and all the other resources we mentioned above. Our process, of course, is the integration of all these resources to convert raw materials into the finished product.

SYSTEM

What resource management really takes is a system—which can be defined as a group of steps or processes organized in such a way as to perform a desired function and yield the desired output. This adds the element of quantity to the inputs and outputs. We need to measure the rate at which we need the output, and correspondingly, set the rate of input to equal that output rate. We can all see what would happen if the input rate were unequal to the output rate.

Let's look at a service example. If an airfreight forwarder took more packages in during the day than it could deliver during the day, the number of packages in the system would build up. If that pile of packages got high enough, and if it stayed in the system long enough, customers would become dissatisfied and the company would lose business.

The same thing is true in a manufacturing company. If the input resources exceed the output demands, the pile of inputs grows larger and larger. In most companies, that pile of excess inputs takes the form of inventory.

The implication is that we need to measure the rates of input and output and compare them; we also need to be able to adjust them to meet varying conditions. This is what management is all about.

Successful management depends on having sufficient understanding of how the process works in the conversion of input to output. A system includes all three: input, process, and output. Everything we will talk about from here on will help us to gain a clearer understanding of how this system works. No matter what kind of business enterprise we work with, *people* are the foundation for its success, as well as one of the resources or inputs. This is illustrated in Figure 1–2.

This view of a system can have multiple meanings, depending on how we look at the business enterprise. The system can be the entire enterprise—the total company—or it can be an individual work center—a specific machine and a specific person performing a specific task. Here, the input would be some raw material, a person, and probably some electricity. The process is the interval during which we add the value to the raw material that makes it into an output. This could be an act such as drilling a hole in a part, or plating the barrel of a pen.

FIGURE 1–2
System (expanded view)

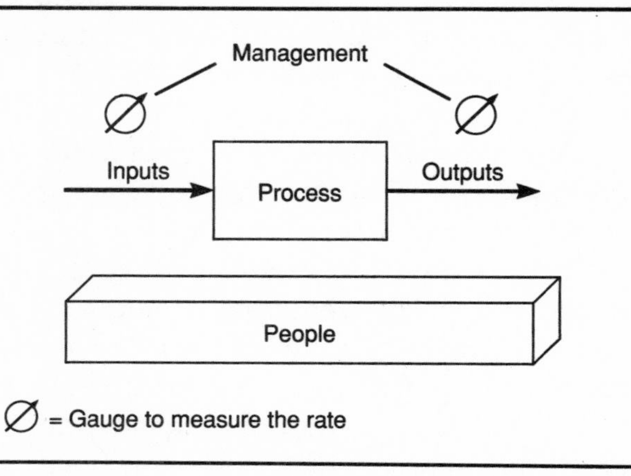

At the other end of the spectrum, our whole company's outputs would be all of the pens, pencils, packaged desk sets, refill leads, and other products that our customers want. Whatever kind of a system we are talking about, we can study it as though we were looking at it through a microscope, where we can adjust the focus to see different levels of detail operating together. Throughout all the rest of this book, we will be examining the tools and techniques that we use to process the resources to achieve the outputs desired. Each of those tools can be viewed as a system just like the one in Figure 1–2.

MATERIAL REQUIREMENTS PLANNING

Nowhere is this truer than in the case of MRP. There are two definitions for MRP. One is material requirements planning, and the second is manufacturing resource planning, commonly known as MRP II.

Material requirements planning, sometimes called "little MRP," is defined in the *APICS Dictionary* as a set of techniques that uses bills of material, inventory data, and the master production schedule to calculate requirements for materials. It makes recommendations to release replenishment orders for material. Further, since it is time-phased, it makes recommendations to reschedule open orders when due dates and need dates are not in phase. Originally seen as merely a better way to order inventory, today it is primarily a scheduling technique; that is, a method for establishing and maintaining valid due dates on orders. This set of techniques is usually a computer program that converts three main inputs—bill of material, inventory data, and the master production schedule—into two main outputs—recommendations for replenishment orders and reschedule notices. This is illustrated in Figure 1–3. This sounds very complicated now, but later, in Chapter 3, when we look at the basic logic of the MRP process, it will become much clearer.

At the other end of the spectrum is manufacturing resource planning, or MRP II. The *APICS Dictionary* defines this as a method for the effective planning of all resources of a manufacturing company. Ideally, it addresses operational planning in units, financial planning in dollars, and has a simulation capability to answer what-if questions. It is made up of a variety of functions, each linked together: business planning, production

FIGURE 1–3
Material Requirements Planning

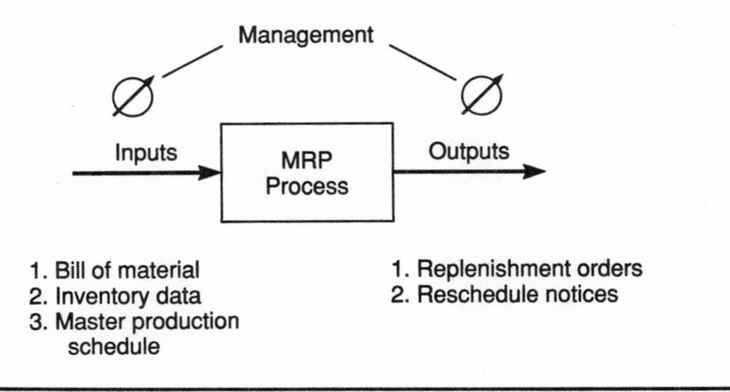

planning, master production scheduling, material requirements planning, capacity requirements planning, and the execution support systems for capacity and material. Output from these systems would be integrated with financial reports such as the business plan, purchase commitment report, shipping report, inventory projections in dollars, and so on. As you can see, manufacturing resource planning takes a much more global view of the business enterprise. So, when people say, "We have an MRP system," it can be hard to tell whether they mean little MRP, the mathematical technique, or MRP II, which plans *all* the company resources.

Throughout this text, we will use the term *MRP system* to talk about both of these concepts, and which of the two we intend can be inferred from the context. Think of it as hearing friends tell you they're from New York. Do they mean New York State, New York City, or the borough of Manhattan? We can ask them to narrow it down for us, or we can pick up from the context of our conversation just what they mean. Usually, "MRP system" will be closer to the concept of MRP II—manufacturing resource planning—with its implications of integration with all aspects of the business.

INVENTORY ORDERING

An MRP system, with its inputs and outputs, starts as a way to order inventory. Let us define inventory as those items that are used to support production and customer service. It can also be expanded to include supporting activities such as maintenance and operating supplies. Right now, we are concerned primarily with the items that serve to decouple successive operations in the process of manufacturing a product and distributing that product to the customer. Inventory can be thought of as the flow of parts from the input through the process to the output.

We can take a closer look at this concept by analyzing a process with its inputs and outputs. First, there has to be some demand for our output, our product. In terms of the whole company, we're looking at our customers' demands for finished goods. The demand comes from our customers primarily, but there may be other demands also, such as some distribution system, or other plants that require our product. Think like our customer for a minute: What is your attitude toward our inventory, and in particular, our ability to deliver products? How soon do you, the customer, want delivery? How does yesterday sound? The base concept is *quick*. Instant delivery is often called just-in-time (JIT) delivery.

But on the other hand, look at our inputs to the process, or where the supply comes from. We have our suppliers, or vendors, who provide the goods, services, and raw materials we need, bringing them in from the outside. Another source of supply is our manufacturing organization, which provides manufactured items. How do these people view the rate at which these things are delivered to our system? How long do they think it should take to provide the things we need? Normally, the term *lead time* comes up in the conversation. The lead time should be long enough to promote an efficient manufacturing operation.

This series of relationships is illustrated in Figure 1–4. You can see that inventory is a holding area where we compensate for fluctuations between the rate of supply and the rate of

FIGURE 1–4
Inventory Management

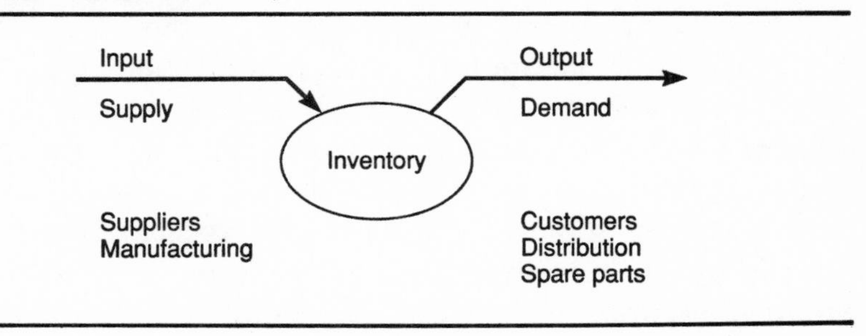

demand. It acts as a buffer in that it decouples the demand rate and the supply rate. In most business environments, the demand rate is very quick when compared to the supply rate, which wants to operate more slowly to take advantage of efficiencies of scale. If we are able to make our supply rate respond very quickly, with small quantities delivered frequently, we can fill customers' demands with very little, or no, inventory.

Here at FineLine Industries, the demand for our product comes from retail stores who buy pens and pencils, including major distributors and nationwide chain stores. Other demand might come from large corporations wanting pens bearing their company logo for use in incentive or marketing programs. Our supply, on the other hand, comes from companies furnishing raw materials such as steel, steel tubing, erasers, and lead. Plus, we have our own manufacturing organization, which converts the raw steel into the components of pens and pencils and then takes those same components and assembles them into finished units.

This leads into another classification of inventory, work in process, or WIP. Manufacturing is in a unique position in that it is on both sides—input and output. The raw materials come in and are converted into components that may be stored in inventory until they are required for the assembly process, at which time the components are drawn from stores, assembled, and shipped to the customer. On the supply side of the fence, manufacturing wants time to build a part; but on the demand side, when manufacturing needs the part to build an assembly, they want it right away.

If we examine the corporate goals of our business organization, as shown in Figure 1–5, we see that our marketing department is primarily concerned with being very responsive to customers' wants. We should be able to make quick delivery of the exact configuration of our product that the customer has ordered. Marketing management also is interested in a high degree of customer service, including good product information, a competitive price, and high-quality products. After the initial customer sale, the marketing department wants to provide

FIGURE 1–5
Corporate Goals (often considered conflicting)

- Marketing management
  - Customer responsive
  - Customer service
- Manufacturing management
  - Efficient use of people
  - Utilization of facilities
- Financial management
  - Minimum inventory
  - Minimum investment

postsale services such as spare parts, repairs, and aftersale maintenance. This implies that we should have sufficient inventory levels to supply anything in our catalog or any spare parts our customers need.

Manufacturing management, on the other hand, is concerned mostly with the efficient use of people and a high utilization of equipment and facilities. This usually means keeping people working, and we have found that they work more efficiently on large-run orders.

But financial management wants minimal inventory and a minimum investment in all of the various input resources. A minimum investment, after all, leads to maximized profits.

You can see that marketing wants to have a lot of everything, manufacturing wants a lot of the things that are easy to make, and finance doesn't want very much of any of it. In many companies, these various viewpoints or goals could be considered to be conflicting. What we need is a planning system that can help people balance these conflicting objectives.

PLANNING

In the planning process, we define the resources we need in order to fulfill the customers' demand. There are many inputs into this planning process, but the three main ones are ex-

FIGURE 1–6
Inventory Management System (planning process)

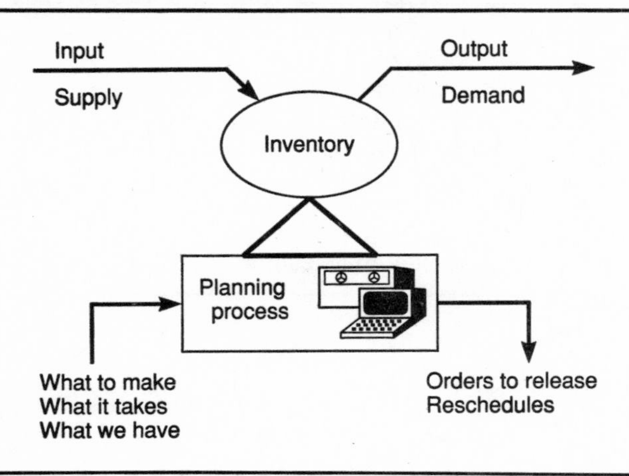

pressed as questions: What are we going to make? What does it take to make it? What do we already have?

If we can answer these three questions, we can determine what orders we need to place and what changes we need to make in schedules that we have already set. We can see this illustrated in Figure 1–6.

EXECUTION

In the real world, there's a big difference between planning and actually getting the job done, or the process known as *execution*. Here, we apply the available resources to produce as planned, so the inputs to our execution process are the resources known as the Four Ms: manpower, machinery, material, and money. The outputs of the execution process are the actual finished goods that are sold to customers. This all comes together as shown in Figure 1–7. The center element, the planning process with its three inputs and two outputs, illustrates material re-

FIGURE 1–7
Inventory Management System (execution process)

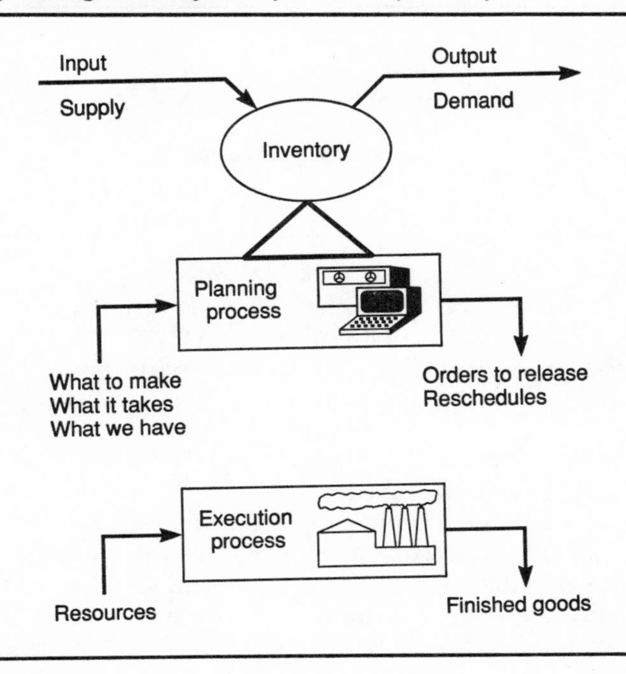

quirements planning—little MRP—with the emphasis on that word *planning*.

Manufacturing resource planning, MRP II, is the more global concept that incorporates the whole process of planning and execution plus control, as shown in Figure 1–8. Our objective is to put techniques into place that are integrated, formal methods of planning and controlling the execution process. Integrated means that all the various elements work together, so that if the customer changes demand, we have the option of changing our plan or not, consistent with our inventory objectives. Formal, as opposed to informal, implies that only one such plan is in place in our company, and we're all looking at the same information expressed in the same way, through a common data source.

FIGURE 1–8
Management System (Integrated, formal methods)

- Plan
- Execute
- Control
  - Set tolerances
  - Measure performance
  - Feedback
  - Corrective action

Whenever we think of an informal system, we're reminded of Otto, the supervisor of the assembly department where Terry had his first real manufacturing job. Terry remembers:

> Every morning, I had to walk through the assembly shop in order to get to the steel shop where I worked, and Otto decided to take me, a new, green youngster fresh out of school, and teach me how a manufacturing operation really worked. He'd take me over to a corner of the assembly department and say, "Look at this pile of material here. Do you know why we haven't assembled it and shipped it out to the customer yet?" "No," I'd say. "Well," he'd continue, "we're missing part number 107-x, and it's tied up in *your* steel shop." So I'd write that number down in my little black book that I carried in my pocket, and I'd go down to the steel shop and ask the production dispatcher which part was supposed to be manufactured next. And what do you suppose the odds were that the dispatcher would name the same part that Otto had just told me about?
>
> Remember, the dispatcher was working within the formal system, but Otto (with me and my little black book) was operating using the informal system. Only Otto and I, and the little black book, possessed the "real" information about what part should be run, but the formal system, the computer, was telling everybody what part needed to be run.

And what system should we be using to run the parts—the formal computer system, or the informal system, where Otto gets his parts? While we can run a company by the informal system, and get the products—some of them—out the door, the formal system makes sure that we're all working together to get the entire plan accomplished. That's why integrated, formal methods are so important in planning and execution.

An element that we use to tie our planning and execution systems together is control methods. Here we find out how close we actually come to our plan. In planning, we set a target —say, the number of pieces we will produce each day. In control, we set a tolerance, which is the amount of deviation acceptable from customer requirements, and measure to see how close we came. If our measurements show that we exceed the tolerance, we need to provide feedback in order to determine the cause of our problems, take corrective action, and bring our planning and execution back into balance. This can be shown in Figure 1–9.

FIGURE 1–9
Control

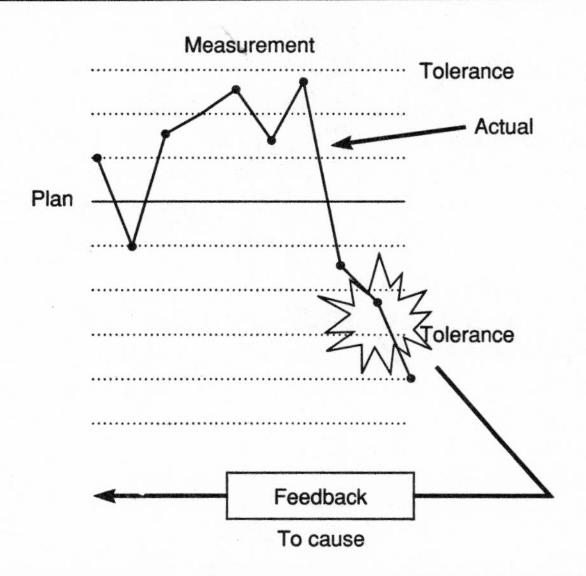

FIGURE 1–10
Business Objectives

- Improve customer service
- Effective manufacturing production
- Minimum inventory investment
- All to improve return on investment (ROI)

BUSINESS OBJECTIVES

The purpose of this entire system is to fulfill our three main business objectives: Improve customer service, make our manufacturing production more effective, and maintain a minimum inventory investment. These three objectives, when realized, will improve our return on investment (ROI), as seen in Figure 1–10.

Improved customer service mostly concerns the speed in which the right product goes to the customer. Note the difference between *efficient* and *effective*. Efficient manufacturing production is making things at high speed. Effective manufacturing production is making the *right* things at high speed. It's possible to be efficient, and make a thing quickly, but make the wrong thing. So *effective* includes the concept of making the right thing, at the right time, at the right speed. And with minimum inventory investment, we're doing all of that with the lowest expenditure of our resources that we can.

There are literally hundreds of techniques that can be used to make this system work. Some of the ways in which they are tied together are illustrated in the closed-loop manufacturing diagram in Appendix D.

CHAPTER 2

INVENTORY PLANNING

Now we turn our attention to defining a planning system that will simultaneously improve customer service, support manufacturing effectiveness, and still maintain a minimum inventory investment. Of course, this system is an MRP system.

Our first consideration must be customer service. And what is customer service? It is, simply, the delivery of the product to the customer at the time the customer wants it. In today's competitive market, we have to remember that the product must be a high-quality item that meets or exceeds the customer's needs. This need for a particular product or component is called *demand*.

DEMAND TYPES

Basically, there are two types of demand: *dependent demand* and *independent demand*. Dependent demand occurs when the need for a product is triggered by some specific event. In manufacturing, this event is typically the requirement for an assembly that uses this product. Independent demand, on the other hand, occurs when the demand for the product is not related to other products or events. An example of this would be a customer's order for end-unit products from our plant.

Let's imagine that you are the materials manager here at FineLine Industries, and you're going to the plant early Saturday morning to catch up on paperwork. It's about 20 miles from your house to the plant, and you're in the habit of stopping at a little roadside diner where they serve homemade sausage and biscuits and the best coffee in three counties. Usually, when you walk in, there are about 60 sausage-and-biscuit sandwiches on the warming table (in inventory). One morning, you ask the short-order cook how she knows how many they need, and she tells you that they usually sell 60. "So how come you knew that I was coming in to buy my usual two?" you ask. "Well," she answers, "it's a busy highway. If it wasn't you, it would be somebody else. That's just how it is." This is a classic characteristic of independent demand for a product; a lot of customers each buy a comparatively small quantity of product, compared to our inventory level. We don't always know exactly who the customers will be, but day after day, the total remains about the same.

One Saturday morning, when you walk in, the cook's cousin is there with her, and they're frying sausage patties and wrapping sandwiches as fast as they can. "What's all this?" you ask, and they explain that the leader of the local Boy Scout troop telephoned yesterday evening. The troop is on its way to a week-long jamboree in Kentucky, and they want to stop by on the bus to pick up breakfast to go for 40 Scouts. And this, you reflect, is a classic case of dependent demand, where a specific relationship exists between an event and customer demand. The increased volume of customers—a busload of hungry Boy Scouts—has caused a jump in the normal sales rate of sausage-and-biscuit sandwiches.

And on yet another Saturday, you walk in and there are sausage-and-biscuit sandwiches wrapped and piled all over the counter. You count 15 piles of 20 each—300 sandwiches for sale. "What's going on?" you ask. "Why so many, and it's not even a workday?" The cook stares at you for a minute and starts to laugh. "Have you forgotten what day it is? This is the first day of deer season. Every hunter going down this road is going to want breakfast." This is a kind of middle ground, a seasonal fluctuation, which is a forecasting situation that demonstrates characteristics of dependent demand in that we can look toward a specific triggering event.

The underlying principle is one that any materials manager knows: Independent demand can be calculated, or *forecasted,* using averages over time. Dependent demand, on the other hand, is calculated with the knowledge of a specific event that is going to inflate usage.

Going back to the diner, we can see that the independent usage of 60 sandwiches every morning indicates that the cook should make 60 biscuits and fry up 60 sausage patties first thing every morning. And when the Scout leader telephones with a requirement that wasn't expected or usual, a large batch has to be ready to meet that demand. The first morning of deer season is an example of forecasted demand with a seasonal increase, so that the cook knows well in advance of the day that she needs to make 300 biscuits and 300 sausage patties.

In any case, however, the quantity of sausage patties needed is dependent on the total number of sandwiches that are planned. To carry this into a manufacturing environment, we can conclude that the demand for a component is dependent on the demand for an assembly.

The diner, in addition to the sandwiches, also serves biscuits on the side. Therefore, there is an independent demand for biscuits alone, making the total demand for biscuits the sum of the dependent and the independent demands, as shown in Figure 2–1.

And what does this have to do with us here at FineLine Industries? In our pen-and-pencil line, there are several exam-

FIGURE 2–1
Demand Types

	Dependent	
Independent	*Sausages*	*Biscuits*
60 sausage-and-biscuit sandwiches	60	60
30 biscuits alone	—	30
Total required	60	90

ples of dependent and independent demand. The sale of pen-and-pencil sets is an example of independent demand. We know that certain seasons—Christmas and June graduations—cause a jump in demand as people buy presents. This demonstrates some dependent characteristics, just the same as deer season at the diner; but overall, the demand for our end units by our customers is independent demand.

We also sell single pens, not in sets, another case of independent demand. So the total demand for pens must be calculated from the total pens needed in sets and as single sales. Going a step further, the demand for a component part of the pen is dependent on, or derived from, the requirement for pens. We need one ink cartridge for each pen we manufacture (dependent demand), and we also need enough ink cartridges to sell as refills for previously sold pens (independent demand). Refill ink cartridges purchased to replace dried-out cartridges in our competition's ball-point pens are another case of independent demand. This is summarized in Figure 2–2.

The point we must remember is that in order to have an effective control system, we must take into account the various sources of demand for our product and its component parts. We will discover that the different demand types are best handled by different types of inventory planning.

FIGURE 2–2
Demand Type Characteristics

Independent	*Dependent*
Definition: Demand that is not related to any other event	Definition: Demand that is related to or derived from another item
Forecasted averages	Calculated from need
Examples: Pen-and-pencil sets Single pens Finished goods Spare parts	Examples: Single pens Ink cartridges Component parts Raw material

INDEPENDENT DEMAND

First, we will focus on independent demand, which is unrelated to specific events and can be forecasted using average usage over time. One effective inventory planning tool to help us manage these independent demand items is the *reorder point*.

Here's how it works: Looking at the quantity of a particular part we keep on hand over a period of time, we would go out to the stockroom and count to see how many we have today. Assuming we use at least one of these parts every day, if we count them again tomorrow, there will be fewer than there are today. Each day, if we repeat the physical counting, the quantity will decrease. If we chart these daily stock counts as points on a graph, we get a line showing us the *rate of demand* for the average usage, as shown in Figure 2–3. The demand, this line tells us, is smooth, uniform, and continuous. So if we extend

FIGURE 2–3
Sawtooth Pattern of Inventory

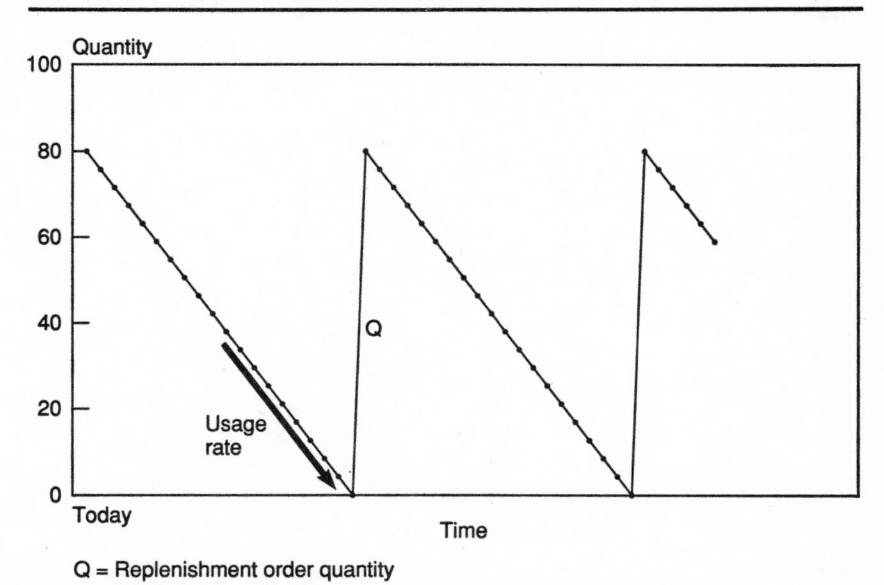

this line into the future, the trick is to predict the point at which we will run out of stock. At that date, we will want a new supply, or a *replenishment order quantity,* to come into the stockroom and bring the balance on hand back up. Day by day, the stock level will fall predictably as it did before. This is commonly called the *sawtooth* pattern of inventory.

In order to have the replenishment order quantity arrive at the time our stock runs out, we need to anticipate the release of the replenishment order. This anticipation time is called *lead time,* the amount of time between the recognition of the need for an order and the receipt of the goods. This creates the first element of the reorder point, which is the demand during the lead time, as we see in Figure 2–4.

But in the real world, we know that any number of things can complicate this process. To begin with, the demand isn't precisely smooth, uniform, and continuous; there are fluctuations, however slight, in the rate of demand. And sometimes

FIGURE 2–4
Sawtooth Pattern of Inventory (lead time)

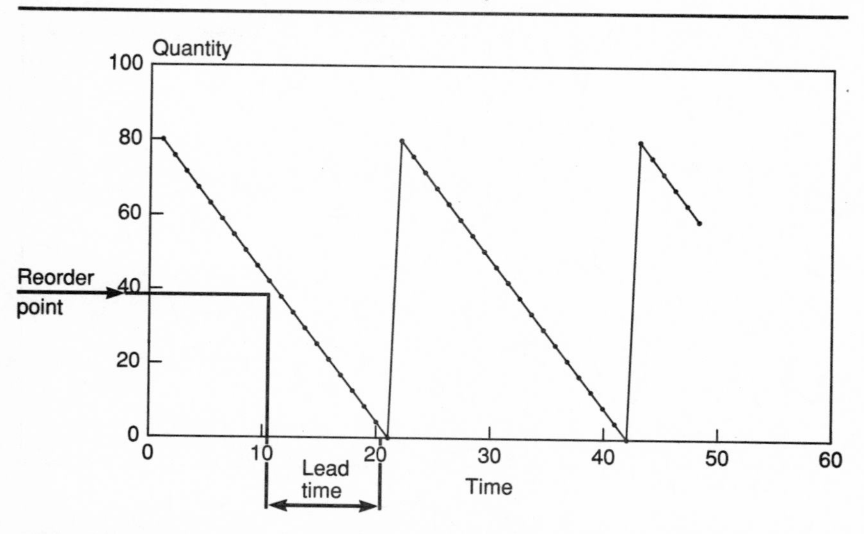

there are fluctuations in the lead time; it takes longer for the material to arrive than we thought it would.

How do we protect ourselves from these fluctuations? There's no reason why the replenishment order has to arrive exactly as the stock level hits zero. We can allow ourselves a cushion, or *safety stock,* as shown in Figure 2–5. This safety stock allows us a margin to deal with fluctuations in the rate of demand or the rate of supply.

Techniques to calculate safety stock quantities are covered extensively elsewhere, and the exact amount isn't really important to this example. For the time being, we merely need to know that the reorder point is the demand during the lead time plus the safety stock that we allow ourselves against contingency. The most important thing to remember is that the reorder point method of planning is most applicable when demand is relatively smooth, uniform, and continuous.

FIGURE 2–5
Sawtooth Pattern of Inventory (safety stock)

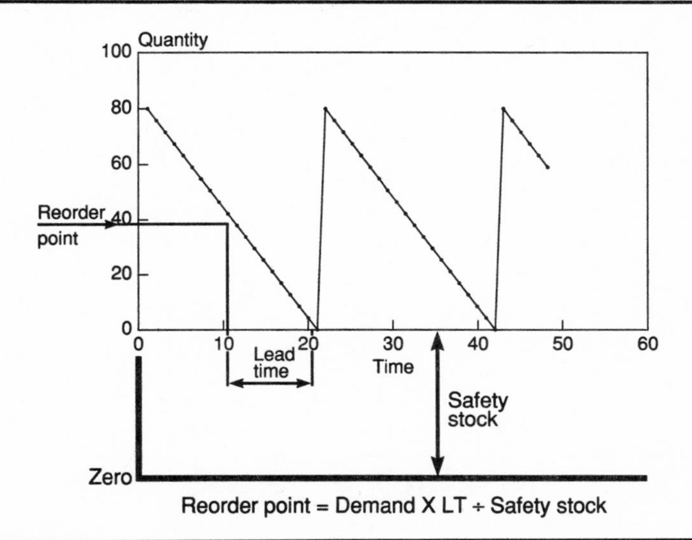

DEPENDENT DEMAND

In dependent demand, however, we have already seen that large fluctuations do occur. To understand why these wide swings occur, let's go back to our pen example where our sales of finished pens have been following a smooth, uniform, and continuous pattern. Until we release the replenishment order to trigger the assembly of more pens, the demand for ink cartridges is going to be at a very low level. But when that assembly order is released, we're going to need a whole bunch of ink cartridges all at once. The stock level will fall off sharply, as indicated by the demand pattern shown in Figure 2–6. The appropriate method to use in inventory planning for our ink cartridges is to anticipate the next "lump" in the demand. If we can predict the release date of the next replenishment order for pens, we can anticipate the quantity and date of the next de-

FIGURE 2–6
Ink Cartridges Demand Pattern

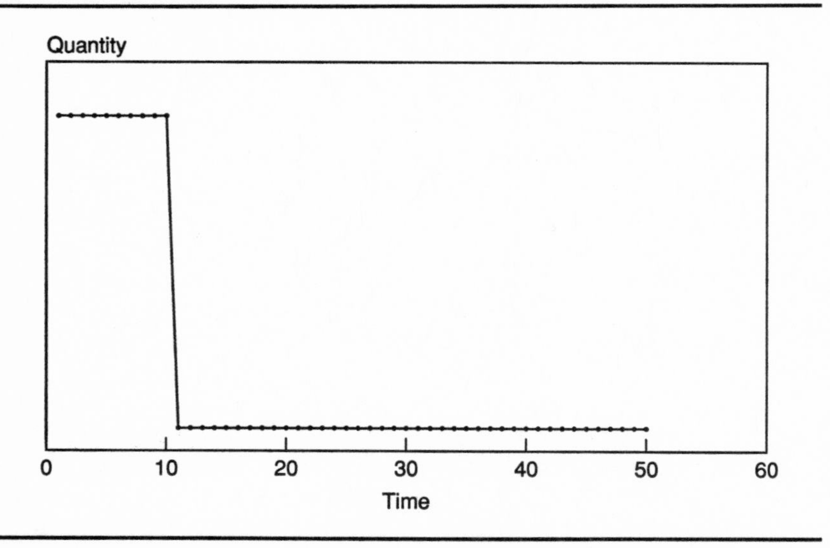

mand for ink cartridges. This gives us the ability to predict the point at which we must place the replenishment order for ink cartridges. (See Figure 2–7.)

When the demand for an item can be calculated through its dependence on another item, as in the case of the ink cartridges, the best technique to use is material requirements planning.

As you can see, MRP looks forward, anticipating future orders based on sharp fluctuations in demand, whereas the reorder point uses historical averages and triggers one order at a time. (See Figure 2–8.) It is important to recognize this difference when we select the appropriate inventory planning technique for each of the many items in our plant. We should focus on the characteristics of the demand for each item. (See Focus 2–A.)

FIGURE 2–7
Dependent Demand (ink cartridges)

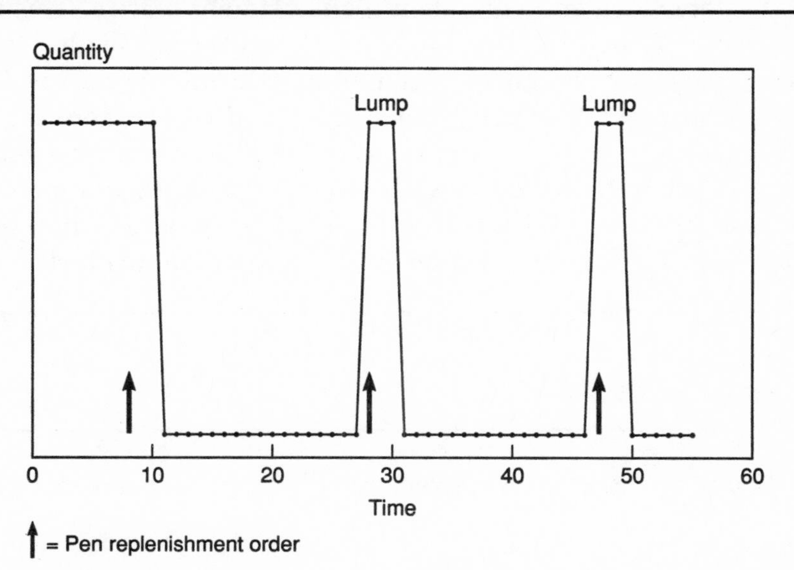

FIGURE 2–8
Two Systems of Planning Inventory

Reorder Point	*MRP*
Looks back	Looks forward
Uses averages	Deals with lumpy demand
One order at a time based on expected runout	Plans multiple orders based on need dates
Smooth, uniform, and continuous	Sharply fluctuating demand

APPLICATIONS

In today's company using independent demand items, the reorder point technique is often used to control the finished goods inventory. Then it is important to focus our attention on a very responsive customer service policy and a comprehensive order entry system. If we can anticipate the rate of demand well enough, then we can reduce the replenishment order quantity and safety stock and still retain high customer service. This is not accomplished by spending a lot of effort in calculating the average usage over time, but rather by reducing the replenishment lead time. A strategy that works effectively in, for example, a medical supply company, would be to increase the safety stocks during the period in which we are stabilizing our manufacturing process and reducing our lead times. When the lead time is shortened, we are then able to respond more quickly to

FOCUS 2–A
Focus for Success

Focus on the *nature of the demand* to use the proper inventory planning process:

Independent—Reorder point
Dependent—MRP

changes in the demand. At this point, we can reduce the replenishment order quantity, approaching just-in-time delivery without sacrificing customer service.

If we look at the control of dependent items through MRP, we see that it is necessary to have responsive production and capacity control systems. This means we need to have the ability to make our capacity flexible—sometimes termed "flexing" capacity—so that we can build components and subassemblies at the exact point when they are needed. In order to manufacture parts quickly, we need cross-trained employees who can perform many different operations, and perhaps we might want to flex the size of our work force. An example would be a service industry, such as an airfreight delivery company, where packages must be processed from one point to another. The company might flex its work force in such a way that, no matter how many deliveries need to be handled on a given day, each package reaches its destination in accordance with customer expectation. This is done by anticipating the demand, or load of packages, early enough in the day to call in the number of employees required to complete the delivery schedule and service the customers. In a service environment, capacity is dependent on customer demands.

Manufacturing companies begin by controlling their parts through the reorder point technique. Gradually, they recognize that some of these components have dependent demand and they evolve into an MRP technique to control the dependent items more effectively. Over time, they discover that parts that have demands derived from many sources create an abundance of future demand. The MRP system generates an enormous amount of paperwork for these parts. When the demand for a particular component approaches uniformity, it is possible to use the reorder point technique for this component, giving us the advantage of reducing the MRP paperwork and treating the component part as if it were an independent demand item. You will recall from our discussion of independent demand items that we can move toward just-in-time delivery by reducing the

FOCUS 2–B
Focus for Success

Use:
MRP for *planning*
JIT for *execution*

replenishment order quantity. This lends itself, in many cases, to a simple visual control system. Companies do this with commonly used components such as paint, greases, resistors, and steel. This is excellent, but we must be alert to changes, either in the design or usage of the finished product, that might affect the usage of the component. We want to continue using MRP logic to assist us in detecting pending changes in usage. Therefore, companies using simple JIT signals for execution will still use MRP for planning. (See Focus 2–B.)

CHAPTER 3

BASIC LOGIC OF MRP

The basic logic of MRP is a simple process that begins with the customer and how well we are able to supply his demands. To illustrate this, let's go back to FineLine Industries, where you are the materials manager. The sales manager has just burst into your office to announce that she's sold some pens. What information do you need? How many pens she's sold, to begin with, and when the customer wants them. Then you're going to want some specifics: Which pen are we talking about? How are they to be packaged? And, in practical terms, who is the customer, so we can start thinking about the proper priority to give the order?

For our purposes, in the basic logic of MRP, we need to know how many pens—let's use 20 for our example—and when the customer wants them—let's say next Friday. Of course, the main question you have as materials manager is whether or not we can fulfill our customer's requirements. The first thing that we should do is to check inventory to see if we already have the finished pens in stock ready to ship. If we do, then we would merely enter a shipping order and not have to get into our manufacturing planning logic.

MRP'S INPUTS

However, let's say that no, we don't have any stock on hand. Where do we go from here? First, we have to ask whether we have the components we need to make the pens, which are illustrated in Figure 3–1. There are three pen parts—the upper barrel, the ink cartridge, and the lower barrel. We need 20 pens

FIGURE 3–1
Pen Components

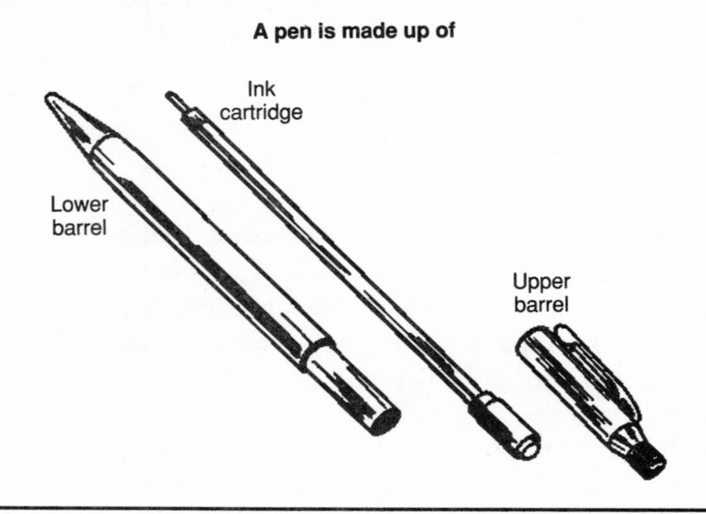

next Friday for this customer, so how many upper barrels do we need? The answer, naturally, is 20, because we need 1 upper barrel per pen. Likewise, for the other components, we will need 20 ink cartridges and 20 lower barrels in order to satisfy the customer requirement.

No matter what you have heard or seen about MRP in the past, what we have just examined is the central logic of an MRP system. We have just encountered the three main *inputs* to that MRP system: What do we need to fill the customer demands? In our case, 20 pens next Friday. This is called the *master schedule.* Next, we check our inventory balances, and if we don't have the parts, we consult the bill of material and determine the quantity of each component required. If we consider all the needs for each particular component, this is called the *gross requirement.* Throughout this first example, we are going to be looking at the requirement for upper barrels, and we know that we need 20 of them by next Friday.

Once this requirement is established, we subtract our inventory from it. We check the stockroom to see how many upper barrels are on the shelf. The stockroom tells us that we have 25 upper barrels on the shelf. Are we capable of delivering the pens by next Friday? Well, it depends on the available stock for the other two components, but we do have enough upper barrels to do the job. If we have 25 on hand, and we need to use 20 to fill the customer order, how many will we have left on the shelf next Friday? The answer, of course, is 5, as shown in Figure 3–2.

What we have just done is commonly referred to as computing the *projected balance on hand*. Our current balance on hand is 25 and we can project the balance on hand next Friday to be 5. There is another element of our inventory information, however, that must be taken into account: Do we have any scheduled receipts for these upper barrels? A *scheduled receipt* is an order for material to replenish our stock. For example, if we purchase the upper barrel from a vendor, we might have an open purchase order with that vendor for a quantity to be delivered on a specific date; many companies call this a scheduled receipt. The term does have a broader implication, for if the part were manufactured in our own plant, we would call it an

FIGURE 3–2
Upper Barrel (projected balance on hand)

	Now	Next Friday	
Gross requirements		20	
Projected balance on hand	25	5	

FIGURE 3–3
Upper Barrel (scheduled receipts)

	Now	Next Friday	
Gross requirements		20	
Scheduled receipts			
Projected balance on hand	25	5	

open order. On our charts, we will portray both of these in the line titled Scheduled receipts, as shown in Figure 3–3.

Gross requirements reduce our projected balance on hand, whereas scheduled receipts increase the expected balance on hand. We should not limit our view to the period between now and next Friday; we should consider conditions that will occur further into the future. Therefore, we will look at our requirements, scheduled receipts, and the projected balance on hand over time. This can be called a *time-phased format*. In Figure 3–4, each column represents a period of time. In our example, each of these periods, or time buckets, represents one week. This is not a necessary condition; not only can the period be something other than a week, the period length can vary across the chart. We will go into more detail about this later when we talk about control parameters.

Back to our example: We have checked our inventory records, and we have a scheduled receipt for 25 upper barrels coming in on Week 2. We enter this number into column 2 of our chart in Figure 3–4. We also want to look at our gross requirements over time. We already know about our requirement for 20 pens in Week 1; other master schedule information tells us that there are 20 more needed in Week 4. Of course, like many other

FIGURE 3–4
Upper Barrel (time-phased format)

	Now	1	2	→
Gross requirements		20		
Scheduled receipts			25	
Projected balance on hand	25	5	30	

companies, we do make things other than this one pen. We also make pencils, as illustrated in Figure 3–5. While the pencil has a number of components that we'll look at later, right now we are going to focus on the upper barrel. If we were to take an upper barrel for the pencil, do you suppose it would fit on the top of a

FIGURE 3–5
Pencil Components

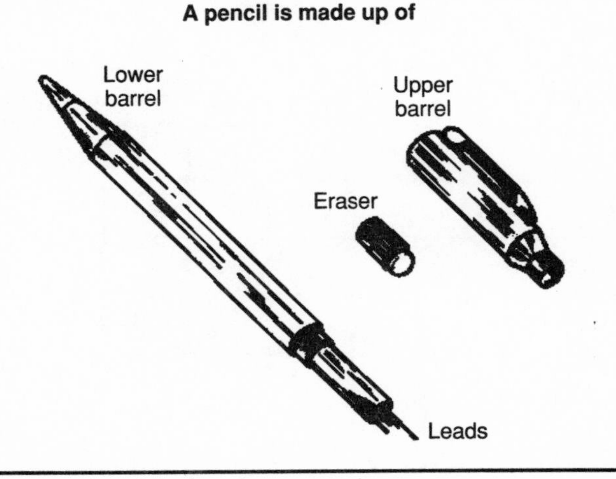

pen? Our engineering department has designed these so that the upper barrels are interchangeable. Consequently, the requirements for upper barrels must include not only the requirements for our pens, but also for our pencils. We know about two pending orders for pencils: 20 pencils for Week 3, and 5 pencils in Week 6. Gross requirements means, simply, that we add together all of the requirements for upper barrels for pens, pencils, plus any other demands. All these requirements are then inserted into our time-phased format under the appropriate dates, as shown in Figure 3–6.

With this data, we can easily compute the projected balance on hand for each week across our time-phased format. The scheduled receipt in Week 2 will raise our balance on hand from 5 to 30, and the gross requirement in Week 3 will reduce the projected balance on hand from 30 to 10. Looking at our chart, we can anticipate what's going to happen in Week 4. If we don't do something by then, our projected balance on hand will go into a negative condition. This means a shortage, and we will be unable to fulfill the requirement for the upper barrel because we will be 10 pieces short. This shortage is defined as a *net requirement* of 10.

FIGURE 3–6
MRP Gross-to-Net Logic (projected on hand)

Upper barrel		Period					
	Now	1	2	3	4	5	6
Gross requirements		20		20	20		5
Scheduled receipts			25				
Projected on hand	25	5	30	10			

FIGURE 3–7
MRP Gross-to-Net Logic (net requirements)

Upper barrel		Period					
	Now	1	2	3	4	5	6
Gross requirements		20		20	20		5
Scheduled receipts			25				
Projected on hand	25	5	30	10			
Net requirements					10		5

Net requirements refer to the quantity that we need in order to fulfill demands. We would show the net requirement for the part each week: In our example, there is a net requirement for 10 pieces in Week 4 and 5 pieces in Week 6, which we show in Figure 3–7. Net requirements mean that we must ensure that more upper barrels arrive into our stockroom so that we can satisfy our gross requirements. How do we do this? We order more upper barrels. This is often called replenishing stock, or a *replenishment stock order*.

How many should we order at a time? The determination of this order quantity is known as *lot sizing*. For purposes of our example, we will set the lot size at 25. This can be determined by the length of the setup time in our shop, or perhaps by the packaging limitations of our vendor.

Logically, we are saying at this point that we are planning to have a receipt of 25 upper barrels come into the stockroom in Week 4. This is called a *planned order receipt*. This represents the quantity anticipated for a specific need date. All this is shown in Figure 3–8.

FIGURE 3–8
MRP Gross-to-Net Logic (planned order receipt)

Upper barrel

	Now	1	2	3	4	5	6
Gross requirements		20		20	20		5
Scheduled receipts			25				
Projected on hand	25	5	30	10			
Net requirements					10		5
Lot size = 25 Planned order receipt					25		

You should keep in mind, incidently, that the difference between a planned order receipt and a scheduled receipt is that the scheduled receipt represents an open order, while the planned order receipt has not yet been released to the shop floor or to the supplier.

Assuming that this planned order receipt will arrive in stock as planned, we can compute an amount called the *projected available*. The projected available takes the balance on hand plus any scheduled receipts minus the gross requirements plus any planned order receipts and records the resulting quantity for each week. This is shown in Figure 3–9.

In order to have this planned order receipt come about, it is necessary to enter an order to replenish our stock. Our planned order receipt, using the lot size, tells us that the order should be 25 pieces, and that it is needed in Week 4. Our next question, then, is, when shall we start this order so that it arrives in the stockroom in Week 4? The length of time from the start of the

FIGURE 3–9
MRP Gross-to-Net Logic (projected available)

Upper barrel	Now	1	2	3	4	5	6
Gross requirements		20		20	20		5
Scheduled receipts			25				
Projected on hand	25	5	30	10			
Net requirements					10		5
Lot size = 25 Planned order receipt					25		
Projected available		5	30	10	15	15	10

order to our finish date is called the *lead time offset*. In most plants, this is the time between the date when the order is released and the receipt of the goods in the stockroom. If we make the part, it is called *manufacturing lead time*, and if we get the part from a vendor, it is known as *purchasing lead time*. In our example, we define our lead time as three weeks. Therefore, we compute that we should release the order in Week 1 in order for it to arrive into the stockroom by Week 4. This is shown in Figure 3–10. This is one of the key elements in the logic of an MRP system: The planned order receipt must be timed to meet the net requirement. We must release this order with a due date that matches the requirement need date. We use the time-phased format to enable us to look far enough into the future to create planned order releases that will get parts on their way into our stockroom soon enough to avert shortages. (See

FIGURE 3–10
MRP Gross-to-Net Logic (planned order receipt)

Upper barrel — Period

	Now	1	2	3	4	5	6
Gross requirements		20		20	20		5
Scheduled receipts			25				
Projected on hand	25	5	30	10			
Net requirements					10		5
Lot size = 25 Planned order receipt					25		
Projected available		5	30	10	15	15	10
Lead time = 3 Planned order release		25					

Focus 3–A.) Also, we should look at any items whose due dates do not match the need date. MRP logic will suggest that we reschedule. In our example, in Figure 3–6, we showed a scheduled receipt in Week 2, but the part is not actually needed until Week 3. MRP would suggest a reschedule notice to move that receipt out to Week 3.

FOCUS 3–A
Focus for Success

Focus far enough into the future
To create *planned order releases* soon enough
To get parts into place
To avert shortages

What we really need to do is condense this information into a format that brings out the key items and shows us what actions we need to take. We don't need to see all of the detailed computations our computer is going to make as it produces its suggestions; we do need to see the results. There are four specific things we need to see: gross *requirement, scheduled receipts,* projected balance *on hand,* and *planned order releases.* This condensed chart is Figure 3–11.

In Figure 3–11, you will notice that the projected balance on hand number assumes that the planned order releases will arrive as anticipated in Weeks 4 and 8. There are other valid methods used to calculate the projected on hand that do not take the planned order releases into account, as shown in Figure 3–12. This shows the projected balance on hand going negative. Because we think that the balance on hand figures should incorporate all of the current information available to us, in this book we choose not to use the negative balance method. We are assuming that the planned order releases will, in fact, arrive as planned. Hence, we will show the balance on hand being increased by planned order receipts, and our subsequent discussions will reflect this.

FIGURE 3–11
MRP Worksheet (upper barrel)

		Period								
	Now	1	2	3	4	5	6	7	8	9
Requirement		20		20	20		5		20	
Scheduled receipts			25							
On hand	25	5	30	10	15	15	10	10	15	15
Planned order releases		25				25				

FIGURE 3–12
MRP Worksheet (negative balance method)

	Now	Period 1	2	3	4	5	6	7	8	9
Requirement		20		20	20		5		20	
Scheduled receipts			25							
On hand	25	5	30	10	−10	−10	−15	−15	−35	−35
Planned order releases		25				25				

If the upper barrel were a purchased part, the MRP worksheet in Figure 3–11 would show us the action necessary to fulfill all our requirements. Looking at the planned order release in Week 1 tells us we should now release a purchase order to our vendor to have the 25 upper barrels come into our stockroom in Week 4. We should also consider delaying our existing scheduled receipt of 25 pieces, due in Week 2, for one week, into Week 3. Also, looking ahead, we will plan another order release for Week 5 with delivery scheduled for Week 8. We might consider a world-class strategy, where we would advise our vendor *now* of this future projected need.

On the other hand, if we manufacture upper barrels in our own plant, we need to proceed with this planning into the level of the components necessary to build the upper barrel. Our upper barrel is comprised of two components: the clip and the sleeve. MRP logic tells us that the planned order releases at one level create the gross requirement at the next level. Therefore, our planned order release for 25 upper barrels starting in Week 1 gives us a requirement for 25 clips that have to be in our stockroom in Week 1. There will be a second requirement for clips in Week 5. There is a corresponding set of requirements for the sleeves, as shown in Figure 3–13.

FIGURE 3–13
Explosion

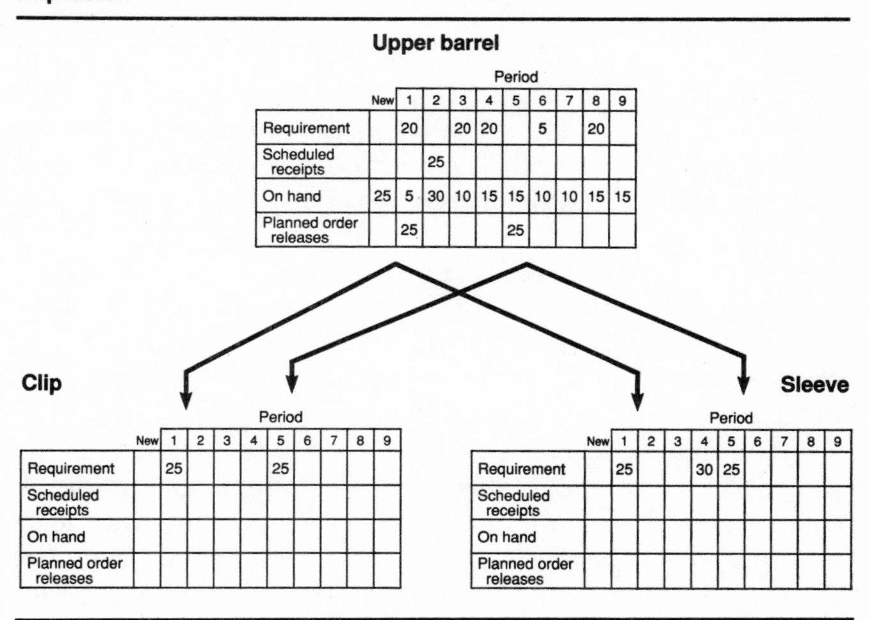

Upper barrel

	Period New	1	2	3	4	5	6	7	8	9
Requirement		20		20	20		5		20	
Scheduled receipts			25							
On hand	25	5	30	10	15	15	10	10	15	15
Planned order releases		25				25				

Clip

	Period New	1	2	3	4	5	6	7	8	9
Requirement		25				25				
Scheduled receipts										
On hand										
Planned order releases										

Sleeve

	Period New	1	2	3	4	5	6	7	8	9
Requirement		25			30	25				
Scheduled receipts										
On hand										
Planned order releases										

We must emphasize that the gross requirements reflect the sum of all the needs for the particular part. In Figure 3–13, you see the requirement for 30 more sleeves in Week 4 that must come from the planned order release for some other kind of upper barrel assembly than the one we have been discussing. It is important that we consider all the requirements for a part before computing the net requirements and making the planned order release. (See Focus 3–B.) Some of these require-

FOCUS 3–B
Focus for Success

Make sure that the *gross requirements* for a part include *all* the demands for that part

ments may be independent demand seen as spare parts, which must be added to the dependent requirements.

The same logic continues for the clips, considering the gross requirements, balance on hand, and scheduled receipts to create planned order releases for the clips. The lot size for the clips may create a planned order release indicating the gross requirement for steel coil. The purpose of this is to ensure that we have sufficient raw materials on hand to manufacture enough clips.

This process of using planned order releases to calculate gross requirements may continue on down through the bill of material for many levels, until we arrive at the purchase level for every part needed in the manufacture of our product. This process is known as an *explosion*.

You may have noticed that the scheduled receipt for the upper barrels in Week 2 does not show a requirement for the components. This is due to the assumption in the MRP logic that the components have already been issued from the stockroom to an existing shop order. The balance on hand for that component was already reduced when the parts were physically moved from the stockroom to the shop floor. This can become more complex, and we will examine the subject in our chapter on net change, when we discuss allocation.

We can distill all of the elements of the MRP logic into an easily understood format, as listed in Figure 3–14. Starting with gross requirements, we subtract inventory, yielding net requirements. To the net requirements, we apply lot size and lead time offset to arrive at planned order releases. Planned order releases at one level create gross requirements at the next level.

Using MRP logic, we can see the three main outputs are, first, planned order releases, used to release new orders either to our vendor or to our own shop. A second output, the reschedule notice, is used to suggest adjustments to scheduled receipts, matching the due date to the need date. Our third output is the data we will use to determine the amount of capacity required to implement our production schedules. We

FIGURE 3–14
Material Requirements Planning (gross-to-net logic)

1. Gross requirements
2. Minus inventory
   Scheduled receipts
   Balance on hand
3. Gives net requirements
4. Apply
   Lot size
   Lead time offset
5. To get planned order releases
6. Explosion
   Planned order releases—at one level
   Gets gross requirements—at the next level

will examine these outputs in greater detail in Chapters 7, 8, and 9.

Still following this MRP logic, we can see that the gross requirements are derived initially from the master production schedule, which is one of the three main inputs into the MRP system. The second input is the inventory status, which we used to determine our balance on hand and our record of scheduled receipts. The third input, the bill of material, is used to convert planned order releases for assemblies into gross requirements for components.

SECTION 2

INPUTS

CHAPTER 4

BILL OF MATERIAL

The first input to the MRP we will discuss is the bill of material. A *bill of material* (or *BOM*) simply refers to the list of items that it takes to make a part. As we were discussing in our pen-and-pencil example in Chapter 3, the bill of material says that the pen is made up of the upper barrel, lower barrel, and ink cartridge.

PART NUMBERING

In order for the MRP process to work, each of the individual parts must have a unique identification. Consider, for example, the ink cartridge. We could have a blue ink cartridge, a black ink cartridge, or any of a number of other colors. Therefore, we must have a unique identifier to distinguish the blue ink cartridge from the other colors. That unique identifier is called a *part number* or *item number*.

It is absolutely essential that each part number be unique and identify one, and only one, specific object. Therefore, when we look at the pen assembly itself, we must have a different part number for the pen with the red cartridge than for the pen with the black cartridge. Of course, the pen would have a different part number from the pencil, because they are two discrete objects. Engineers call this a "difference in fit, form, or function."

There are several different ways of defining part numbers: *random*, *significant*, and *semisignificant*. A random part number is a number that is used as an identifier only, not as a descriptor.

On the other hand, a significant part number tells us specific information about the item, such as source, material, shape, and description.

As an example of a significant part number, we could define an ink cartridge to be 37-1-3-16-432, where each set of digits tells us something about the part. (See Figure 4–1.) The advantage of this system is to help people to recognize the item by merely looking at the part number. The disadvantage of the method is that the slightest change in the physical characteristics of the part will force us to change the part number. The rule should be as follows: If the form, fit, or function does not change, the parts are interchangeable and therefore, the part number should not change.

The worst feature of this method is if we introduce another variable, such as the option of plastic or metal tubing, not only do we need to introduce a new part number for the new ink cartridge, we must also add another digit to all the existing part numbers to distinguish the new from the old. (See Figure 4–2.) This renumbering of the old parts to add the new feature option is a costly step.

An alternative to the constant proliferation and growth of part numbers is to be totally random and to issue part numbers in sequence each time engineering encounters the need for a new part number. The advantage of this is that there is no growth in complexity, and it is easy to do with a computer

FIGURE 4–1
Significant Part Number (ink cartridge)

Part number: 37-1-3-16-432

Type of item:	37 = Ink cartridges
Type:	1 = Screw-in type
Point type:	3 = Fine line
Color:	16 = Blue
Length:	432 = 4.5 inches long

FIGURE 4–2
Ink Cartridge (add on)

Part number: 37-1-3-16-432-3

Type of item:	37 = Ink cartridges
Type:	1 = Screw-in type
Point type:	3 = Fine line
Color:	16 = Blue
Length:	432 = 4.5 inches long

Add new digit

Tube material:	3 = Metal

system. The disadvantage of the method is that there is no practical way for people to confirm that the part number they are recording matches the item they are handling.

To achieve some of the advantages gained when people can relate to the part numbers, a compromise is the semisignificant part number, in which the first segment of the part number tells us what the item is, with the balance of the number being randomly assigned, as shown in Figure 4–3. For FineLine Industries, the part number classification table is shown in Figure 4–4.

The part number should be used by shop personnel to move the part through the manufacturing process and by the computer on our bill of material.

FIGURE 4–3
Ink Cartridge (random part number)

Part number: 37-7213

Type of item:	37 = Ink cartridges
Four digit:	7213 = Random

FIGURE 4–4
Part Number (classes)

Finished Goods		*Components*		*Raw Materials*	
Ink cartridges	37	WIP	20	Steel tubing	10
Pens	42	Purchased complete	22	Flat steel	12
Penciles	43			Lead	16
Sets	51			Eraser	18
Desk sets	62			Packing	19
Spare parts	83				

PARENTS AND COMPONENTS

The key to understanding the bill of material list is the concept known as parent/component relationship. *Components* are the objects that are assembled together to make a *parent*. In our pen example, the pen is a parent with three components: upper barrel, lower barrel, and ink cartridge. Notice that each of the components can, in turn, be a parent. Our upper barrel is made up of a sleeve and a clip, as we see in the bill of material in Figure 4–5. The component called a clip is also a parent, made up of the raw material steel.

FIGURE 4–5
Upper Barrel Parent-Component Relationship

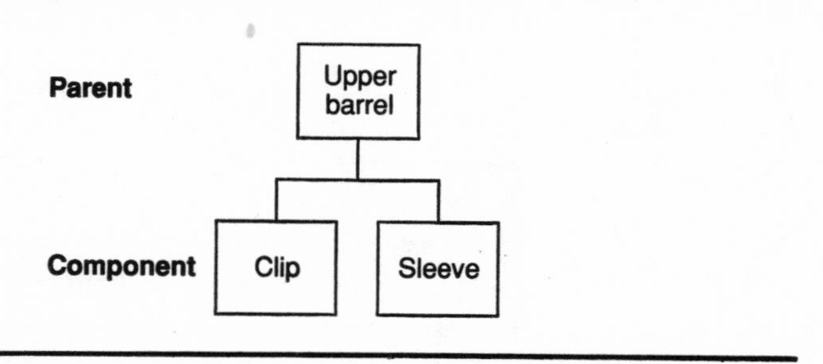

There are several data elements necessary to accurately describe this parent/component relationship. First is the parent part number; next, the component part number; and third, the quantity of the component needed to build one of the parent. In the case of the pen, there are three such parent/component relationships. Many companies today add another data element, called the *sequence number*, to help define the sequence in which these three records should be listed. Such a listing is called a single-level bill of material, as in Figure 4–6.

One of the more significant things to watch for in the development of an MRP system is the required quantity for each component. The quantity in the bill of material should reflect the number of pieces of the component that it takes to make that assembly. In our pen, it is obvious that there is one of each of the components needed to make a pen. Some authorities advocate making allowances for scrap within these quantity designations. For example, if, during the pen assembly, the line workers drop 5 percent of the ink cartridges on the floor, making them unusable, we might increase the quantity per pen to 1.05 cartridges. A more effective method is to keep the quantity per pen equal to the exact requirement and to add an additional data element called a *scrap factor*. The MRP process will combine the quantity per unit plus the scrap factor in order to calculate the gross requirement. This keeps the standard cost accurate and allows us to review scrap factors to point us toward process improvements.

FIGURE 4–6
Single-Level Bill of Material

Parent 42-208: Pen, gold, blue ink; 1 each

Component	*Sequence*	*Quantity Per*	*Description*
20-231	010	1.000 each	Lower assembly, gold pen
37-7213	020	1.000 each	Ink cartridge, blue
20-201	030	1.000 each	Upper barrel, gold

BILL OF MATERIAL LEVELS

We know that the single-level bill of material shows each parent/component relationship, and that each component may, of itself, be a parent, as shown in Figure 4–7. Here, the upper barrel (Part 20-201) is used in both the pen and the pencil. Remember that the upper barrel has a bill of material of its own, which we saw in Figure 4–5. The upper barrel has its own unique inventory data, no matter where it is used. The reason for this is that we must consolidate the gross requirements for the upper barrel before we can calculate the MRP planned order.

Once the parent/component linkages are established, we can display them linked together forming a *hierarchy* (or tree), as in Figure 4–8. We can take any product and go down through the levels of the bill of material all the way to the purchased items. This is called a *multilevel bill of material*.

When we examine a multilevel bill of material tree such as for pencils in Figure 4–9, we can see that a component can

FIGURE 4–7
Single-Level Bill of Material

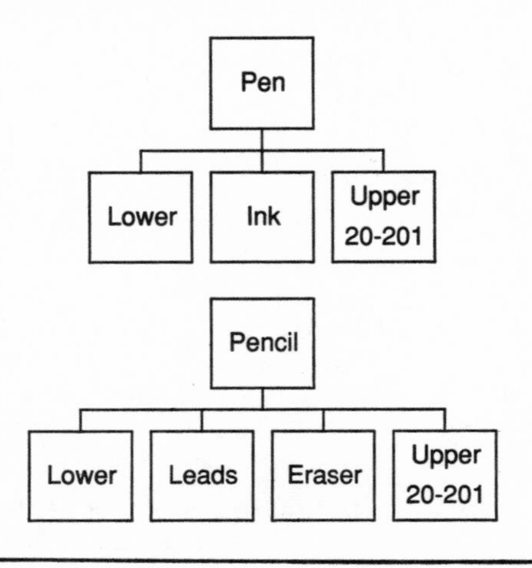

FIGURE 4–8
Multilevel Bill of Material Tree for Pens

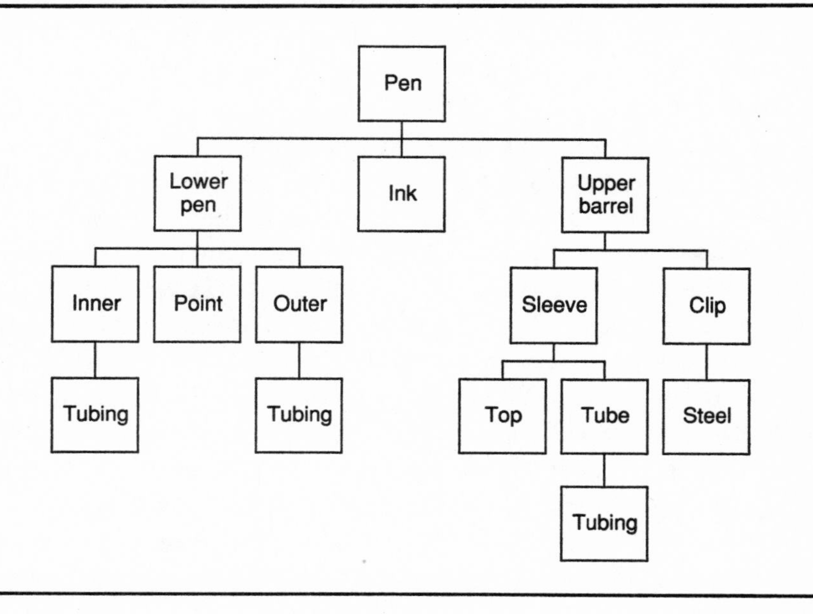

FIGURE 4–9
Multilevel Bill of Material Tree for Pencils

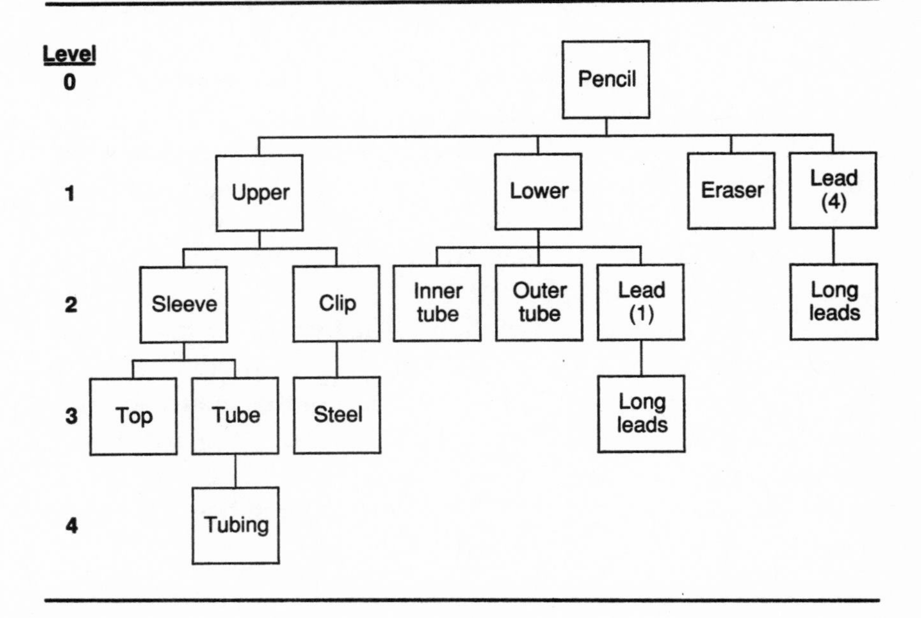

appear in more than one place within a bill of material. Notice that the lead appears as part of the pencil assembly because we need four pieces there. We also need one piece of lead to be inserted into the lower barrel subassembly. Accordingly, the lead appears in two different levels of the pencil bill of material.

By convention, we usually number the levels of the bill of material starting with Level 0, representing the end unit in the BOM. The pencil is Level 0, the upper barrel is Level 1, the lower barrel inner tube is Level 2, and so forth. The lead appears both in Level 1 and in Level 2. The MRP logic is that the gross requirement must be the sum of all requirements, meaning that we cannot net out the inventory of the leads until the MRP has progressed to at least Level 2 in this bill of material. Therefore, we must put a code in the lead item to show how low it appears in *any* bill of material. This is called the *low-level code,* and will be mentioned again when we talk about inventory data in Chapter 5.

When discussing bills of material for parts we manufacture, we should remember one additional important piece of information: the *routing*. This is defined as the sequence of steps that it takes to manufacture the product. It should show the work center (the machine or work area) where each operation is to be performed, and the sequence of the operations. It should also include setup, run-time standards, processing time, and perhaps tooling and other information that will help operators build the product. Combining the bill of material with the routing gives us the ability to visualize what it takes, in both materials and labor, to make the part.

We can print this information in the form of a *manufacturing shop order,* a document that can be given to the production shop to tell them how to produce the part. Part of this manufacturing order is typically called the *kitting list,* or the *picking list,* which gives the total numbers of each part needed to complete the order. We will see how our stockroom people use this list in Chapter 7.

EXPLOSIONS AND IMPLOSIONS

Once we have all of this data in one database, we can display it in several different ways, which fall into two basic categories: *explosions* and *implosions*. Explosions go down the bill of material, whereas implosions go up the bill of material. A *single-level explosion* is the same as a single-level bill of material, as shown in Figure 4–10. Second, an *indented BOM explosion* shows a multilevel bill of material, with the parent on the left-hand side and each additional level indented farther to the right, as in Figure 4–11. A *summarized explosion* is nothing more than rearranging an indented explosion into part number order and adding together the total requirement for each part number, as in Figure 4–12. Notice that this tells us we need five pieces of lead for a pencil. Explosions are typically used by material planners to assist them in determining lower-level component availability.

Implosions, on the other hand, follow the same concept as explosions, but in reverse. A *single-level implosion* is often called a *"where used" list*. Figure 4–13 shows us that an upper barrel is used in both the pencil and the pen. An *indented implosion* goes on to the next level, which shows that the upper barrel is used in a pencil, which is then used in a set. (See Figure 4–14.) A summarized implosion rearranges an indented implosion in

FIGURE 4–10
Single-Level Explosion

Parent 43-208: Pencil, gold, thin lead; 1 each

Component	*Sequence*	*Quantity Per*	*Description*
20-241	010	1.000 each	Lower subassembly, gold pencil
20-201	020	1.000 each	Upper barrel, gold
18-108	030	1.000 each	Eraser assembly
16-108	040	4.000 each	Lead, thin 3/4″ long

FIGURE 4–11
Indented BOM Explosion

Parent 43-208: Pencil, gold, thin lead; 1 each

Level	*Component*	*Quantity Per*	*Description*
1	20-241	1.000 each	Lower subassembly, gold pencil
.2	20-235	1.000 each	Inner sleeve, gold pencil
..3	10-104	0.489 feet	Tubing, inner
.2	20-236	1.000 each	Outer tube, gold pencil
..3	10-103	0.375 feet	Tubing, outer casing
.2	22-431	1.000 each	Point, gold pencil, thin
.2	16-108	1.000 each	Lead, thin, 3/4″ long
..3	16-100	0.066 feet	Long thin lead
1	20-201	1.000 each	Upper barrel, gold
.2	20-429	1.000 each	Outer sleeve, gold
..3	20-213	1.000 each	Upper barrel tube
...4	10-103	0.133 feet	Tubing, outer casing
..3	22-212	1.000 each	Upper barrel top
.2	20-211	1.000 each	Upper barrel clip
..3	12-113	0.003 pound	Steel, spring clips
1	18-108	1.000 each	Eraser assembly
.2	18-109	1.000 each	Eraser 1/2″ long
..3	18-101	0.525 inch	Eraser material 1/4″ diameter
.2	18-110	1.000 each	Eraser socket
..3	12-108	0.001 pound	Steel eraser socket
1	16-108	4.000 each	Lead, thin, 3/4″ long
.2	16-100	0.264 feet	Long thin lead

order by part number so that we can readily identify the end units that use this component, as shown in Figure 4–15.

This implosion information is typically used by engineers to demonstrate the effect of design changes in components. This shows all of the end units and the various levels within the bill of material that will be affected by the change.

In our company, we want to be sure that we merge the bill

FIGURE 4–12
Summarized BOM Explosion

Parent 43-208: Pencil, gold, thin lead; 1 each

Level	*Component*	*Quantity Per*	*Description*
..3	10-103	0.375 feet	Tubing, outer casing
...4	10-103	0.133 feet	Tubing, outer casing
	Total	0.508 feet	
..3	10-104	0.489 feet	Tubing, inner
..3	12-108	0.001 pound	Steel eraser socket
..3	12-113	0.003 pound	Steel, spring clips
.2	16-100	0.264 feet	Long thin lead
..3	16-100	0.066 feet	Long thin lead
	Total	0.330 feet	
1	16-108	4.000 each	Lead, thin, 3/4″ long
.2	16-108	1.000 each	Lead, thin, 3/4″ long
	Total	5.000 each	
..3	18-101	0.525 inch	Eraser material 1/4″ diameter
1	18-108	1.000 each	Eraser assembly
.2	18-109	1.000 each	Eraser 1/2″ long
.2	18-110	1.000 each	Eraser socket
1	20-201	1.000 each	Upper barrel, gold
.2	20-211	1.000 each	Upper barrel clip
..3	20-213	1.000 each	Upper barrel tube
.2	20-235	1.000 each	Inner sleeve, gold pencil
.2	20-236	1.000 each	Outer tube, gold pencil
1	20-241	1.000 each	Lower subassembly, gold pencil
.2	20-429	1.000 each	Outer sleeve, gold
..3	22-212	1.000 each	Upper barrel top
.2	22-431	1.000 each	Point, gold pencil, thin

of material together in one database. Several different departments may utilize several different bills according to their needs. Engineering is concerned with product design, manufacturing is concerned with how the product is built, marketing

FIGURE 4–13
Single-Level Implosion (where used)

Component 20-201: Upper barrel, gold; 1 each

Parent	*Sequence*	*Quantity Per*	*Description*
42-208	010	1.000 each	Pen, gold, blue ink
43-208	010	1.000 each	Pencil, gold, thin lead

FIGURE 4–14
Indented BOM Implosion

Component 20-201: Upper barrel, gold; 1 each

Level	*Parent*	*Quantity Per*	*Description*
1	42-208	1.000 each	Pen, gold, blue ink
.2	51-101	2.000 each	Set, gold pen and pencil
1	43-208	1.000 each	Pencil, gold, thin lead
.2	51-101	2.000 each	Set, gold pen and pencil

FIGURE 4–15
Summarized Implosion

Component 20-201: Upper barrel, gold; 1 each

Level	*Parent*	*Quantity Per*	*Description*
1	42-208	1.000 each	Pen, gold, blue ink
1	43-208	1.000 each	Pencil, gold, thin lead
.2	51-101	2.000 each	Set, gold pen and pencil

with how it is used, and production planning with how it is planned. It is important to merge all these requirements into one common database.

PHANTOM BILLS

There are several types of bills of material that can be used to facilitate a variety of needs at the same time. One of these is called a *phantom bill of material*. This is also called a *transient;* that is, it is never intended to be stocked. Further, we do not want to create planned orders for this transient subassembly, and so we code it a phantom bill, or *blow-through,* where the lead time is defined as zero and the lot size as lot-for-lot.

We can understand these more easily by looking at an example. Our engineering and marketing departments tell us that we are going to start selling pencils with different company logos on the upper barrels. Therefore, they want us to introduce a subassembly part number on the bill of material for all the parts of the lower barrel, consisting of the lower barrel subassembly, the eraser, and the four pieces of extra lead. Out on the manufacturing floor, however, we know that they are not going to build orders of these subassemblies. What they are going to do is take the lower barrel, four pieces of lead, the eraser, and the appropriate upper barrel and assemble them at one time. So engineering's new subassembly will only exist while it is going down the assembly line. This allows MRP logic to blow through the subassembly and show us requirements for its components, and still gives engineering the ability to use the part number and planning to use it only if there happens to be some inventory.

MODULAR BILLS

Another kind of bill that is often used to help us with marketing our products is a *modular bill.* This is usually employed when a product such as an automobile can be sold with a number of different options. Another example might be pens with the option of several different ink cartridge colors. The purpose of a modular bill is to identify the various ways our product can be sold. Normally, we have two separate bills of material, one for the pen with blue ink, one for black. While there may be other colors as well, we will use only these two for simplicity. We could force marketing to forecast each color ink and use our MRP logic to make planned orders. However, marketing tells us that, while they can forecast total pen sales pretty accurately, it is much trickier to try to distinguish color preferences. Our practical response to this dilemma is to create a modular bill, which we will call "Pens," and which is made up of an upper barrel, a lower barrel, a blue cartridge, a black cartridge, and any other color cartridge commonly used. Of course, each pen requires one upper barrel and one lower barrel, and we can use historical averages to estimate the quantity of each color ink cartridge. In Figure 4–16, we have estimated 60 percent blue and 30 percent black. As marketing forecasts the total sale of

FIGURE 4–16
Modular Bill ("Pens")

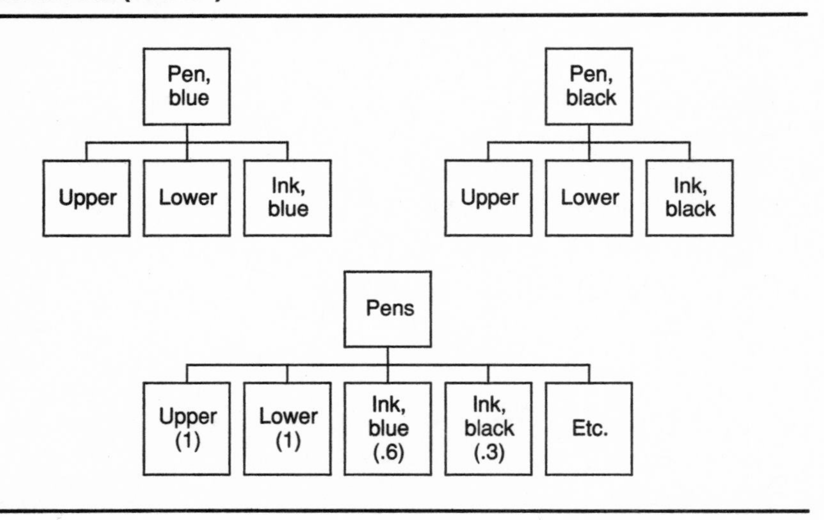

pens, the bill of material explosion will calculate the gross requirement for each component.

Some companies even have software systems that can incorporate this into the order entry system, so that customer service representatives can "configure" various assemblies by choosing the options desired. This type of bill of material can facilitate forecasting and planning. It also gives us the ability to do what is commonly called *overplanning,* in which, if we think that the blue ink cartridges may sell more than their anticipated percentage, we might increase the quantity per to a higher percentage. We will show how these bills are used in Chapter 10. This helps to ensure a high level of customer service while avoiding a mismatch in our finished goods inventory.

PLANNING BILLS

We can carry this one step further and go to a completely artificial grouping of all the options at the top level of the bill of material that represents the total product line. This type of bill is called a *planning bill, superbill* or *pseudobill* of material, and may

FIGURE 4–17
Planning Bill

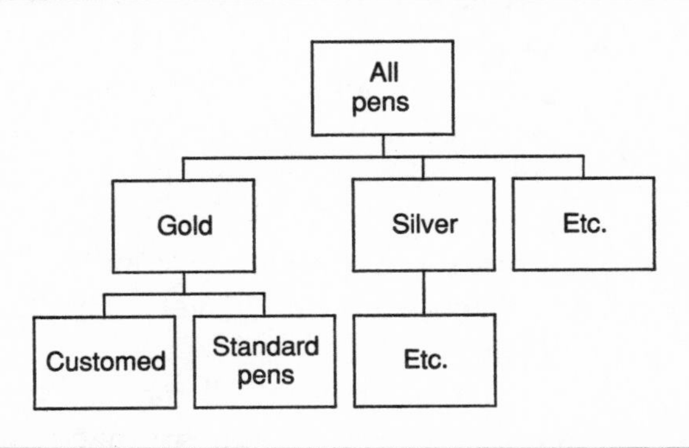

be used for forecasting at an even higher level than our modular bill. (See Figure 4–17.) These bills are used to help facilitate planning the aggregate production plan for the product family and represent an item that is never actually built. Additionally, we could use these bills to assist us in planning the load on the manufacturing resources, so we attach a form of the routing to these planning bills, known as a *bill of labor*.

BILL OF LABOR

A bill of labor shows the work centers, usually only the critical work centers, and can be used to show the average standard rate. We can multiply the forecast by these standard rates to estimate the load in our manufacturing shop. We may also want to measure other resources, such as floor space, electricity, and even supplier requirements. Taken to this extent, it is usually known as a *bill of resources,* giving us the ability to compare the load to our capacity for that product line.

APPLICATIONS

Experienced users of MRP systems have found that the multiple-level processing of bills of material can become cumbersome. Excessive levels that the bill must go through, with the gross-to-net logic and the moving of parts in and out of inventory, can become very costly. Some companies have gone so far as to put a level in the bill of material for each step of the routing. It is superfluous to plan going into or out of the stockroom for each of these steps; therefore, it is best either to code them as phantoms, or to remove them completely from the bill of material in order to reduce the number of levels. This is known as *flattening* the bill of material.

Some companies have removed all manufacturing levels from the bill, so that it reflects their synchronized manufactur-

FIGURE 4–18
Airfreight Planning Bill ("packages")

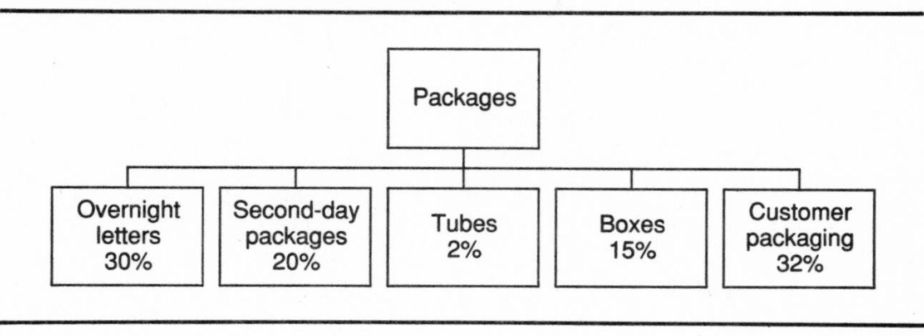

ing process. Therefore, we do MRP and master schedule planning of the end-unit product. The MRP will create planned orders for purchased components. In the just-in-time environment, these planned orders can be communicated to the supplier as an advance warning of our requirement; then some visual pull signal can be used to actually authorize the delivery of the component.

Service industries can also use these bill of material concepts to assist them in planning the utilization of their items. For example, if we are trying to predict the usage of airfreight packaging materials, we may use a planning bill called "Packages" that has the percentage of product mix for overnight letters, second-day packages, tubes, and boxes. (See Figure 4–18.) If our total demand were 1 million, we could then compute that we need 300,000 overnight letters every day. If we are forecasting an increase to 1.2 million, we would then know that we will need 360,000 overnight letters every day with a corresponding increase in the other areas as well. As you can see, the only limit in the application of these basic principles is the imagination.

CHAPTER 5

INVENTORY DATA

There are two basic types of inventory data: *dynamic* and *static.* Dynamic inventory data deals with those types of elements that are in constant change, such as the physical balance on hand, scheduled receipts, and all the other elements of the gross-to-net logic.

On the other hand, static inventory data consists of identifier information such as part number and description. Other elements of static inventory data are those items that we use as control parameters. We shall see that these parameters determine the quantity and date of the planned order releases.

An organized collection of records, whether computerized or manual, is known as a *file,* and specific areas in those records where a particular category of data is recorded are called *fields.*

DYNAMIC DATA ITEMS

First, we will examine the dynamic inventory data items. The first of these is the *physical balance on hand,* or the *quantity on hand,* the quantity shown in the inventory records as being physically in stock. All of our MRP calculations start with the current balance on hand. This is one of the data elements that must be absolutely correct. If there is a difference between the MRP quantity and the balance physically present in the stockroom, all of our subsequent calculations will be wrong. If the physical quantity is less than the computer believes it to be, we will experience shortages and therefore we will be unable to supply customer demands. On the other hand, if the physical balance is larger than the computer balance, our decisions will inflate the inventory. It is imperative that we have a method in place to monitor the accuracy of the on-hand balances. The most effective technique for maintaining this accuracy is cycle counting, which we will discuss in more detail later on.

Closely associated with the physical balance on hand are the records that indicate the storage locations of these balances. In some cases, these may even reside in a separate file, the location file, that has records showing the quantity on hand of a part number stored in a given location. The sum of the quantities of the part in each of its locations should be equal to the total balance on hand. There are two basic methods of storing materials in stockrooms: *fixed* and *random* locations.

Fixed locations mean that a specific part is always stored in the same place in the stockroom. This implies that the stocking location must be large enough to accommodate the maximum replenishment order that could ever be received. This can cause inefficient utilization of the storage areas; to overcome this inefficiency, we choose random storage locations. In this system, we can store items in any available location or locations, but we must have a record-locating system to tell us where we put them. This system must be able to tell us all the separate locations where a particular part is stored, and how many are stored in each location. It should also be able to tell us all of the parts stored in any location, allowing us to verify our record accuracy by location. The sum of all these records is the total balance on hand that is used by the MRP logic.

Another dynamic data item is *reserved materials,* which are materials on hand but earmarked or designated for specific requirements such as a customer or a future production order. In some systems, these can be called *allocated materials.* It is important to know what materials have been designated as belonging to future requirements, so that we can calculate what is known as the *available balance on hand,* sometimes referred to as *available inventory.* This is the on-hand balance less allocations, reservations, back orders, and quantities reserved for quality problem resolution.

All of this deals with the information that is valid as of today's date. Our dynamic inventory data should also track information that can calculate these numbers into the future, or the *projected balance on hand*.

Another important element of dynamic inventory data is the concept of scheduled receipts. We have already discussed the fact that scheduled receipts can be open production orders or open purchase orders. Associated with each of these records will be the part number, the quantity, and the due date. Another concept, used mostly in the purchasing environment, is the *blanket order* with a vendor, a long-term commitment for materials, where portions of the total purchase are released to deliver to us as needed over time. Scheduled receipts and balance on hand are two of the main lines on our MRP worksheet.

The other components of the gross-to-net logic make up the balance of our dynamic data. One of these is gross requirements, the sum of all the needs for a particular part. This will include the sum of both the independent and the dependent demand for the part. As you will recall from Chapter 2, dependent demand is the requirement derived from the demand for other components or end items, while independent demand is the forecast demand for such things as spare parts.

PEGGING

From this data we can create a data element known as *pegging*. The pegging for any given item shows the source (or sources) for the gross requirement. Pegging is useful to the material planner to help resolve problems. While we should always strive to meet schedules, we might want to peg our requirements back to their original source, which will lead us ultimately to the master schedule item that may be affected by a change in the component we are reviewing.

The next dynamic item is the net requirement. We know from the basic logic of MRP that gross requirements, minus the balance on hand and the scheduled receipts, yields the net requirement. For each time period, we may have a net requirement that ultimately requires a stock replenishment quantity. We know that we can apply lot sizes to these net requirements to arrive at the *planned order receipts,* the quantity expected to arrive into stock on the date of the net requirement.

Another element of dynamic data is *planned order releases,* which is nothing more than the date on which a replenishment order should be released. It is determined by subtracting the lead time from the planned order receipt date.

There can be other elements of dynamic data defined by specific software systems, and our list is not meant to be all-inclusive.

We can visualize all of these dynamic data elements, shown in Figure 5–1, making it apparent that there are three separate factors to track for each data element: the part number, the quantity, and the action date. If an action date becomes past due, that is an indicator to cue us to take corrective action. It is

FIGURE 5–1
Dynamic Inventory Data for Each Specific Part

		By Date		
Data Element	*Past Due*	*1*	*2, 3, . . .*	*Total*
Gross requirements				
Dependent				
Independent				
Scheduled receipts				
Production orders				
Purchase orders				
Blanket orders				
Balance on hand				
By location				
Allocated				
Available				
Net requirement				
Planned orders				
Receipts				
Releases				

also useful to visualize a total for each data element, so that as the total changes, the plan for the part number can be re-evaluated.

In Figure 5–2, we list a number of typical data elements and indicate the actions that cause the data to change. We will be discussing these concepts throughout the book.

DATA ACCURACY

When we consider such large quantities of data, the importance of accuracy becomes obvious. The best-known method for assuring this accuracy is the technique of *cycle counting*. This is an inventory accuracy audit technique in which we physically count the quantity on hand in the stockroom. A predetermined number of items is counted each workday, with each item counted at a prescribed frequency, or cycle. We can compare the count with the inventory data record to determine if there is a

FIGURE 5–2
Dynamic Inventory Data

Data Element	*Affected by*
Physical balance on hand	Transactions
Locator file	Moves
Allocations	Orders
Available	System calculated
Scheduled receipts	
Production (shop) orders	Shop work order release
Purchase orders	Purchase order generator
Blanket orders	
Gross requirements	
Dependent	Higher levels
Independent	Forecasted
Net requirements	System calculated
Planned order receipts	Lot sizing
Planned order releases	Lead time offset

discrepancy. The primary purpose of cycle counting is not to correct the record so much as it is to determine the cause of any discrepancy. We should research the transactions that cause error and eliminate those causes. Most people assume that this means firing the person who made the mistake; a much better approach is to activate training programs, simplify the procedures, and provide the tools to make it easy for employees to maintain accurate data. We should look for methods to make procedures fail-safe to the point where it is easy to do something right and hard to do it wrong.

Taking cycle counting beyond the physical inventory, we can apply the principles to all the data we use. Any data elements falling into the Past Due column indicate that action wasn't taken when it should have been. We should take steps to correct the causes of inaction. We can also examine data at future points, to make sure that all the elements are accurate, thus implementing a preventive measure against their contributing to a past due situation. We can check the accuracy of the scheduled receipts in purchase orders, the accuracy of the allocations, and even the accuracy of the gross requirement, which in turn is a measure of the accuracy of the bill of material. Any program that verifies record accuracy must be an ongoing effort. This routine method might be termed *cycle auditing,* implying that our purpose is to verify the integrity of all our data records and to eliminate the causes of any discrepancies. All records must be as near perfect as possible. (See Focus 5–A.)

FOCUS 5–A
Focus for Success

Focus on *data integrity.* All records must be as near perfect as possible.

STATIC DATA ELEMENTS

Now we will examine static data elements, as shown in Figure 5–3. These are the data elements that are unlikely to change over time, and they can be subdivided into two groups: the *identifiers* and the *control parameters*. Identifiers include the part number, which is usually the key to the file. Along with the part number, we need a description of what the part is, and probably

FIGURE 5–3
Static Inventory Data for Each Specific Part

Identifiers

- Part number
- Description
- Unit of measure
- Group technology codes
  - Product class
  - Size
  - Shape
  - Inventory class
- Cube size
- Weight
- Shelf life
- Low-level code
- Engineering drawing
  - ECN
  - ECN date
- Purchased part
  - Purchase unit of measure
  - Conversion
- Cost
  - Standard
  - Actual
- Usage
- Buyer/planner

Control Parameters

- Lead time
- Safety stock
- Source code
  - Control code
- Lot size
- Scrap
- Cycle count
  - Frequency
  - Date last count
  - Number of counts
- Joint order code
- Commodity code
- ABC code
- Order modifiers
  - MIN
  - MAX
  - Batch

some other fields describing the item characteristics such as shape, size, shelf life, or space requirements. These fields are the verbal descriptions that help people understand what kind of part they are dealing with. To more fully describe the part, we should include other elements such as engineering drawings, the most recent engineering change notice (commonly known as an ECN), and the date of that change. For many parts, it is also useful to know what items, if any, can be substituted without affecting the form, fit, or function.

One other very significant file element is the unit of measure, the way in which the use quantity of the item is expressed: each, pounds, gallons, feet, cases. There may be two units of measure—one used to track the item through our inventory system and the other the purchase unit, which may be different. For example, steel structurals may be purchased in pounds and measured in feet. In process industries, an item might be purchased in gallons and used in milliliters. Therefore, when we have a purchase unit of measure in the file, we must include a conversion factor to convert the purchase order amounts into our manufacturing unit of measure.

We also should include a field that identifies the cost. The most common cost stored in the field is the *standard cost*, which includes the material, labor, and overhead charges needed to produce the part. Some companies use costs other than standard, such as average cost, most recent cost, or no cost at all if there are wide retail fluctuations, as is the case with precious metals or some petroleum products.

Other fields in some systems might indicate the usage of the item, either as usage by time period or a single gross usage such as per year. This usage number is used to plan the frequency of deliveries, or even how often to schedule cycle counts.

Another field might be product class, used to group like parts together into a product family. Product classes are usually used to sort output reports and assist us in controlling a product line. Another, similar field, the inventory class code, is used to sort and group together like items for aggregate inventory planning.

Another useful field to help us manage items is the buyer/planner code, where we identify the person who is responsible for decisions regarding the part. While some companies indicate the buyer here, and others indicate the planner, in more progressive environments these two positions are merged into one area of responsibility.

The second type of static data is control parameters. The first of these is lead time. Another, used for independent demand items, is safety stock. These two parameters were discussed in Chapter 2, where we described their use in determining the reorder point.

Another series of fields deals with ordering policies. The first, the *order policy code,* differentiates between purchased items and manufactured items and is sometimes referred to as the *source code.* Other order policy codes tell us whether the item is controlled as a master schedule item, or through the MRP, or through some other management method—for example, subcontracting. We should also look at the method by which we order the items, often referred to as a *joint ordering code.* This is used to group similar items together so that they may be ordered together. This is typically done to improve the utilization of material or to group similar items for setup efficiency. A common example of this is the consideration of major and minor setups on punch presses, where we can set a die in the press that will produce both left- and right-handed parts at the same time. A joint ordering code would be entered for both parts, so that we would get ordering suggestions for both at the same time. We should, however, investigate the source of the demands to make sure that we really need to produce both parts.

Another frequently used code is the *commodity code,* usually used for purchased items. An example might be nuts and bolts, chemicals, or printed products. This is akin to the concept of group technology, where we try to identify the sameness of parts. If we are manufacturing screws, they all have similar geometries, so they should be thought of and planned as a group, rather than as discrete individual parts. This can provide for efficient retrieval of similar designs and anticipates cellular-type production processes.

Another control parameter is the *ABC classification,* which can be used to define cycle count frequency or order frequency. Generally, A items should be more tightly controlled than B items, which in turn are more tightly controlled than C items. The effort saved in less-stringent monitoring of C items can be devoted to the control of A items. Some companies use these classifications in determining order quantities, but a better method is lot sizing.

LOT SIZING

Lot sizing is the process of determining order quantities that will satisfy the net requirements and the economics of supply, which can be either purchase or manufacture. While there are countless ways of determining the appropriate lot size, most authorities agree on nine common methods, as shown in Figure 5–4.

A common method in use today is the *fixed order quantity,* which is really a seat-of-the-pants approach to lot-size determination. In this case, we set a predetermined, fixed quantity, usually based on the physical restrictions of the supply method. These include truckload size or pharmaceutical batches. The resulting inventory level can vary greatly depending on the nature of the demand.

Another method, *lot-for-lot,* produces the exact quantity for the net requirement for each period. This is called a *discrete method,* which means that the quantity produced matches the requirement and there are no remnants left over at the end of the period. The balance will fall to zero. This method will give us the minimum inventory levels and is most appropriate for manufactured components. The disadvantage of this method is that

FIGURE 5–4
Lot Sizing

Technique	*Based On*	*Results*
Fixed	Supply	
Lot-for-lot	Order by order	Minimum inventory
Fixed period	Discrete	Cycles
Economic order quantity (EOQ)	Demand	Remnants
Period order quantity	Discrete	All about the same
Least unit cost	Discrete	
Least total cost	Discrete	
Part period balancing Look ahead Look back	Discrete	
Wagner-Whitin	All possibilities	

we are creating a setup in our shop for each period, whereas it might be more economical to gather some periods together into a single replenishment order. This is called the *fixed period lot size,* which is a discrete method that results in manufacturing cycles for the part.

There is a dichotomy between minimizing setup costs and minimizing inventory levels, and the most popular method for resolving this is the *economic order quantity,* or *EOQ*. The EOQ sets a fixed order quantity that is based on minimizing the sum of those two costs. In order to evaluate these numbers, we must determine an anticipated annual usage quantity. This makes the EOQ a demand-based lot-sizing technique that will always yield remnants. Built into the EOQ is the assumption that usage is smooth, uniform, and continuous, and we know from our basic logic of MRP that the net requirement for components is lumpy. This makes the EOQ a very poor choice for MRP lot sizes.

Over the years, people have developed variations of the EOQ formula in attempts to make it a discrete calculation. (See Figure 5–4.) The results are relatively similar and can be used to make nearly any lot size look better than the others. The common thread through all of these techniques is cost data such as carrying cost and inventory cost; as a whole, these techniques are more applicable to very high-cost items.

For a detailed discussion of the formulas and applications of each lot-sizing technique, see Appendix C.

Once the lot size has been determined, there are still many factors involved in the purchasing/procurement process. These modifiers may be an *order minimum* or an *order maximum*. The minimum is the smallest quantity of the part that should ever be ordered. This could be due to some sort of process constraint, such as a pharmaceutical batch minimum. On the other hand, a maximum order quantity might be appropriate where item size or storage facilities dictate that we should never produce more than the given quantity.

Another modifying field might be designated multiples, in those cases where it is smart to produce an order quantity that might fit on a skid carried by a fork lift, or that will make maximum utilization of raw material cutting. This is sometimes known as the *unit of issue*, or *batch quantity*, depending on the industry.

Another modifier to be applied to the lot-size logic is *scrap*. There are several different ways of defining scrap. One is *yield*, or *process yield*, which means that the replenishment order is started at one quantity, but when the order is completed, the yield is a lesser quantity of usable pieces. To arrive at the planned order release quantity, we need to take the planned order receipt quantity and divide it by 1 minus the scrap factor. To return to our pens, if we experience a 10 percent scrap rate in the crimping operation for the lower barrel, we will need to release 111 pieces in the order to receive 100 pieces into stock. This is illustrated in Figure 5–5.

It is extremely important to get our planned order release quantity correct so that the MRP logic will compute the proper quantity of requirements for components. Our objective is ulti-

FIGURE 5–5
Scrap Fall Off

Planned order receipt	100 pieces
Scrap 10% at crimp	
Planned order release	111 pieces

$$\text{Qty} = \frac{\text{Need}}{1 - \text{Scrap}} = \frac{100}{1 - 0.1} = \frac{100}{.9} = 111$$

You see:	111 pieces start
Less:	11 bad pieces (10% scrap)
	100 good pieces left

mately to reduce or eliminate the cause of the scrap. We can utilize the "scrap" field to stratify the parts and identify the areas needing management attention.

Different industries may call the scrap factor by different names, for example, *shrinkage factor*. In the process industries, the most common term is *yield factor*.

Two other types of scrap involved in the process are *kerf* and *cutting allowances,* most often used in metal industries, though their application can be seen in other manufacturing processes. The crimped lower barrel is 1⅞ inches long. If the tubing is sawed to this length, there is a certain portion of the tubing that is consumed in the sawing process. This is called the kerf allowance. Therefore, we must plan on, say, $1\frac{15}{16}$ inches of raw material to produce each barrel. In addition, we may also need to make a first square cut on the tubing to facilitate feeding the material into the machine, and after the pieces are cut, there may be a remnant length to be held in the machine. Therefore, we might get only 20 pieces out of a 4-foot piece of tubing. We need to allow 2.4 inches of tubing per piece, and to get the 111 pieces we need at the start of the crimp process, we are going to need 22.2 feet of tubing.

Since our tubing only comes in 4-foot lengths, our assumption is that the biggest piece of "dropoff" can be used for subsequent orders, but we are still going to be unable to utilize every inch of every 4-foot piece of tubing.

Some people hide all of these allowances in the Quantity Per of the bill of material, which makes the computation come out right; but a better approach is to put it into the scrap field to give it visibility. This allows us to identify a process improvement step.

The last group of control parameters is maintained by the computer system to help us monitor our activities. Two of these are the *cycle count frequency*, telling us how often the item should be counted in a year's time, and the date of the *last cycle count*. We use these fields to determine when the item should appear on a cycle count list.

Another control parameter maintained by the system is the item's low-level code, the lowest level at which the part appears in any bill of material tree.

These static and dynamic files define the inventory data related to each part. Of course, they need to be linked with the bill of material files that define the parent/component relationships. Manufactured parts would also be linked to a routing file, which describes the steps taken to manufacture the part. This will be examined more fully when we discuss capacity requirements planning.

APPLICATIONS

Companies that are using MRP systems discover two significant parameters that have the most effect on their ability to manage the inventory system. These are the planned lead time and the lot size. Planned lead time is merely that—the time an order *should* take between time of release and time of receipt. Some companies have made these numbers very large, expecting to gain control of their processes. The opposite is usually the result. The longer an order is in process, the harder it is to control. Consequently, we should strive to reduce the lead times to as short as possible. When we get into more detail on capacity planning, we will discuss specific techniques that will help us do this.

There's an old saying that tells us, "Whatever the lot size is, it should be cut in half." Lot size, however, is related to setups. Therefore, if we believe in this concept of halves, our first focus should be cutting the setup in half. Go out onto the shop floor with the production people, or with the vendor, and study the process to see how we can alter it to cut the setup by half. Then we can cut the lot size in half. Cutting our lot size allows us to provide high customer satisfaction with minimal inventory investment, and we can go on to halve our setup again, with a subsequent halving of the lot size.

Another item getting a lot of attention currently is the ABC code. While it was once used only to track inventory, today it is used to isolate the important few (A items) from the trivial many (C items). We see companies using ABC stratification to identify key items for just-in-time (JIT) process improvements. The largest-moving items are the ones we should look at first when we are searching for ways to improve the process. We also see companies using the balance on hand to identify those items that should be the focus of inventory reduction. ABC stratification can be used on open purchase orders to identify candidates for blanket ordering and to better manage accounts payable cash flow.

Process industries will often introduce another parameter called *days supply*, most often used to control end items. This defines the quantity of a finished product targeted to be in the finished goods distribution system. It can be used as an aggregate management of the inventory levels. As we look at process improvements, we may want to increase the level of finished goods in order to maintain excellent customer service while we are changing our process. After we get control of the new

FOCUS 5–B
Focus for Success

Focus and keep:
Lead time short
Lot size small

process, we can use the parameters to draw the finished goods inventory back down as our supply rates become more closely synchronized with our demand rates.

To pull several of these points together, we might want to increase our finished goods inventory while we are experimenting with setup reduction and decreasing our lot sizes. Once we have learned how to reduce lot size and shrink lead times, then we have the manufacturing process more closely aligned with customer demands. (See Focus 5–B.)

Companies that are Department of Defense contractors have other elements that must be tracked closely. Contract specifics may vary, but a frequent area of concern is commingled storage. We can append a part number to another field, the *lot number,* to differentiate one set of parts from a subsequent receipt of the same part. This enables us to demonstrate lot traceability. Beyond this, lot numbers can be used to link certain parts to certain contracts. Once a part is identified in conjunction with a specific contract, it may need to be stored in a unique location.

Another factor that may enter into DOD contracting concerns explosive materials, which may have stringent shelf life requirements. This may call for another field, the *Q/D ratio,* which indicates the quantities that can be stored in juxtaposition and the required safe distance between quantities.

CHAPTER 6

MASTER PRODUCTION SCHEDULE

Simply put, the master production schedule (MPS) is our anticipated build schedule. It is the set of planning numbers identifying the quantity of specific part numbers that can and will be built by the manufacturing firm. This is the schedule that drives the MRP system, and it is one of the three main inputs to the MRP logic.

This build schedule is for those items that are designated as master production schedule parts. They should represent specific configurations; that is, a specific end unit or a high-level component of an end unit. Considering our pens and pencils, we ship the pens loose and we also ship pens as part of the pen-and-pencil set. If we master scheduled end items, we would master schedule the pen-and-pencil set and also the loose pens, so we are actually scheduling two parts. It makes better sense to master schedule just the pens, put them into stock, and use them to build sets on customer demand. This is called master scheduling a high-level component; that is, the pen. Later, when we discuss priority planning control, we will look at two-level master scheduling, which will explain in more detail how to do this.

Once we have defined our specific configuration, we must look at the demand for that part and balance it against the supply for that part. Hence, we will be planning specific quantities to be produced on a specific date. The build schedule, therefore, is the list of the specific dates, quantities, and parts to be built. For this schedule to drive the MRP system, it is necessary that the unit of measure for these quantities match the unit of measure on the item master record. However, it is helpful in many cases, in fostering management teamwork, to convert the unit of measure into a common unit of measure for the entire company, most frequently dollar amounts. Our master production schedule would therefore show the quantity of parts needed for each specific pen, but could just as easily be added up to indicate the manufacturing cost dollar value of the total pens to be produced.

In other company environments, these quantities might be converted into total pounds of steel or some other format that management can relate to easily, because the total master production schedule must be a doable plan. It is best, accordingly, to aggregate the numbers into one common unit of measure.

DEVELOPING THE MASTER PRODUCTION SCHEDULE

First, we should turn our attention to the details of the master production schedule (MPS); then, we will go on to look at the details of the capacity plan. We are going to create a time-phased format that can be used to show the balance between the demand for our part and the supply of the part.

There are two elements of demand for the part, the first being the forecast and the second being the actual customer orders. We will master schedule Pen 42-208 in order to illustrate how the master production schedule is derived.

We begin with the forecast, which typically comes from our marketing department, estimating the quantity of the pen to be sold in each period of our time-phased format. For purposes of this illustration, the forecast will be 10 pens for each period, as shown in Figure 6–1. The second element of the demand is the customer orders that we already have for the product. Usually we would want to list those as the quantity of the part required for each period. Normally we expect that the total of the customer orders would approach the forecast amount in the first

FIGURE 6–1
Master Production Schedule (product demand)

Pen 42-208		Period							
	Now	1 4/23	2 4/30	3 5/7	4 5/14	5 5/21	6 5/28	7 6/3	8 6/10
Forecast		10	10	10	10	10	10	10	10
Actual orders		8	4	2		2			
On hand									
Available-to-promise									
Master schedule									

period and fall off in the periods farther out. We also expect that when we get to those future periods the total demand will come fairly close to the forecast. This concept is illustrated in Figure 6–2.

We must understand how this concept can be expanded to cover the entire family of pens, called the production plan. The production plan for the family shows a level rate of production

FIGURE 6–2
Pen 42-208

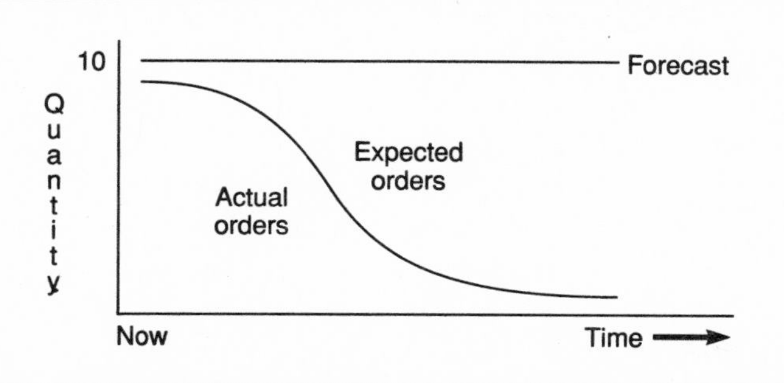

throughout the time period, with the backlog (customer orders received but not yet shipped) representing a large portion of the production in the early time periods. In later periods, the forecast predominates. It is important that the backlog consume the forecast as we book new orders, as we see in Figure 6–3.

To return to the time-phased format, we will now look at the supply of our product, which begins with the current balance on hand of the pen itself. We can project the balance on hand for future time periods in the same way that the MRP logic does. However, there is a question as to whether we should compute the balance on hand based on forecasted sales or on the actual orders. For the purpose of planning the amount to be produced, we should anticipate that the sales will meet the forecast, and so we compute the projected on-hand balance using the forecast line of our chart. At the same time, we use the on-hand balance less the actual customer orders to compute what is known as *available-to-promise.* Available-to-promise is a tool used for customer order promising and represents the uncommitted portion of our inventory. In our example in Figure 6–4, the 30 on hand at the beginning of Period 1, less the 8 customer orders in Period 1, gives us 22 available-to-promise at the end of Period 1.

FIGURE 6–3
Production Plan (all pens)

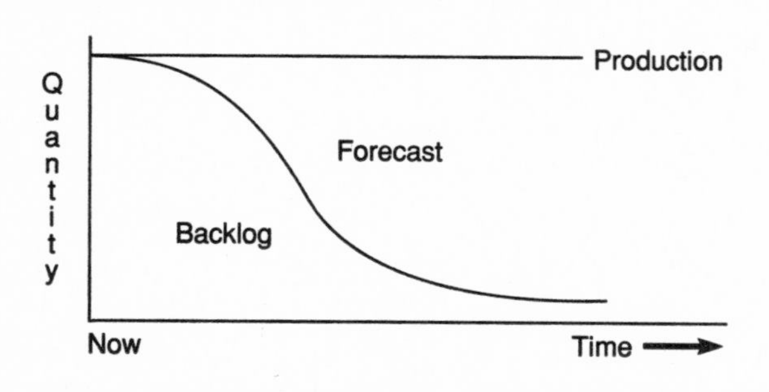

FIGURE 6–4
Master Production Schedule (product supply)

Pen 42-208	Now	Period 1 4/23	2 4/30	3 5/7	4 5/14	5 5/21	6 5/28	7 6/3	8 6/10
Forecast		10	10	10	10	10	10	10	10
Actual orders		8	4	2		2			
On hand	30	20	10	0					
Available-to-promise		22	18	16	16	14			
Master schedule									

Continuing with this logic, the available-to-promise will drop to 14 by the end of Period 5. Right here, the logic of available-to-promise is slightly different from our normal time-phased logic, in that available-to-promise is the *uncommitted* portion of the inventory of the pens, which means that we don't really have 22 pens to promise to a new customer at the end of Period 1, without stealing pens necessary to fill future actual customer orders. Our actual available-to-promise is 14 because we must anticipate the future customer deliveries, as shown in Figure 6–5. The proper logic for available-to-promise should be that it is the *smallest* number in the available-to-promise field prior to the next production run.

This available-to-promise is used by marketing to make future sales commitments; if the number goes negative, we have overcommitted our planned production. On the other hand, the projected balance on hand is used by the master scheduler to project future production. If the projected balance on hand goes negative, as it would in Period 4 in Figure 6–5, it is necessary for us to receive a stock replenishment order for these pens in Period 4.

FIGURE 6–5
Master Production Schedule (actual available-to-promise)

Pen 42-208		Period							
	Now	1 4/23	2 4/30	3 5/7	4 5/14	5 5/21	6 5/28	7 6/3	8 6/10
Forecast		10	10	10	10	10	10	10	10
Actual orders		8	4	2		2			
On hand	30	20	10	0					
Available-to-promise		14	14	14	14	14			
Master schedule									

Since the master production schedule is the "master of all the schedules" and drives the MRP, this is the most important decision-making process in the entire MRP system. Consequently, a person must manage the master production schedules by deciding the quantity to be produced and the timing of that production going into stock. He should understand that he must consider both the material availabilities and the capacity availability. It is also important that the sum of the master schedules of individual parts should equal the total as defined in the production plan.

To illustrate this process, our master scheduler has determined that he must produce 30 pens in Period 4 and 40 pens in Period 7. Figure 6–6 shows the two planned orders. This process sets up a replenishment cycle for this part that takes into account the lead time to manufacture the part, plus the available capacity compared to the forecasted sales. All of these factors must be consistent with the production plan. If, for example, our aggregate production plan for pens stated that we planned to build pens at a rate of 300 pens per period, the master scheduler is responsible for spreading out the production of any one

FIGURE 6–6
Master Production Schedule

Pen 42-208	Now	1 4/23	2 4/30	3 5/7	4 5/14	5 5/21	6 5/28	7 6/3	8 6/10
Forecast		10	10	10	10	10	10	10	10
Actual orders		8	4	2		2			
On hand	30	20	10	0	20	10	0	30	20
Available-to-promise		16	16	16	44	44	44	84	84
Master schedule					30			40	

model of pen in order to accomplish two things at once. First, he must ensure material availability to meet the demand for each individual pen, while second and at the same time, never letting the total production of all pens exceed 300 in any given period. (See Focus 6–A.)

Returning to Figure 6–6, our planned production requires us to recalculate the planned balances and the available-to-promise. The projected balance on hand is computed using the master schedule and the forecast, and so in Period 4, we start

FOCUS 6–A
Focus for Success

A master schedule must—

1. Ensure material is available to meet individual demands
2. Not exceed the production plan

—both at the same time.

with 0 balance on hand, 30 come into stock, we expect to ship 10 to customers, and we are left with a projected on hand of 20 at the end of Period 4. Therefore, we are assuming that the master schedule replenishment orders are coming in at the beginning of the period, allowing us to fill customer demands throughout the time period. We can also reevaluate the available-to-promise. The new production orders, of course, will be available to promise against new customer demands. Additionally, we may also be able to reevaluate the current available-to-promise, because some of our future customer demands may be filled from future production orders. Notice that in Figure 6–6, the customer order for 2 in Period 5 can be taken from the master schedule order of 30 in Period 4. This allows us to increase the available-to-promise in the earlier periods to 16.

CAPACITY PLANNING

We need to assure ourselves that the master production schedule is consistent with our capacity to produce it. We know that the production plan rate is consistent with our resource capability because our production plan is consistent with our resource requirements plan. At FineLine, these plans were all set up to produce an aggregate of 300 pens in each time period. Now, however, we are at the point where we must ensure that the actual product mix will not cause an overload in our manufacturing facility.

For our pens, for example, if the anticipated mix at the production plan level is evenly split between gold and silver pens, our resource requirements plan can be in balance. However, at the master production schedule level, we might mistakenly have one period in which all the pens were gold plated, overloading the gold-plating workstation; consequently, we will be unable to build the master production schedule. We want to take the master production schedule, then, and compare it to our rough-cut capacity plan. This is done through a technique known as the *product load profile.*

The product load profile is a listing of the key resources required to manufacture the part. This includes the critical work centers, or a grouping of operations, no matter where in the bill of material the work occurs. In our example, Pen 42-208 requires 0.1 hours of gold plating and 0.3 hours of machining to produce all of the components used in the pen. Multiplying our master production schedule times this product load profile allows us to predict the capacity required in these work centers in future time periods, as shown in Figure 6–7. We can do this for all of the items on the master schedule to come up with an estimate of the total capacity required in these key work centers. This is called a *rough-cut capacity plan.*

The reason behind this, of course, is that we need to make sure that it is practical for us to produce all of the master schedules we have set. If our rough-cut capacity plan shows an overload in any one particular area, we must resolve this problem by either flexing our capacity or adjusting the master schedule, while still keeping in mind that we must supply product to meet customer demand. If, for example, we notice that we have overloaded the gold-plating operation in Period 7, we may choose to build the master schedule of 40 pens one period earlier, in Period 6, in order to smooth the capacity required. The price for this, of course, is a slight increase in our inventory

FIGURE 6–7
Product Load Profile (Pen 42-208)

Bill of Labor: To build 1 pen requires:
0.1 hours—gold plating
0.3 hours—machining

	Time Period							
	1	2	3	4	5	6	7	8
Pens to Build per MPS				30			40	
Labor hours								
Gold plating				3			4	
Machining				9			12	

in Period 6. We need to perform this problem resolution at the master schedule level, before we run our MRP program.

GUIDELINES TO CONTROL THE MPS

Now we need to consider time fences. This *fence* concept should be considered more as a guideline of time constraints, as opposed to walls that we can't climb over. The first time fence is the *planning horizon*. This is the distance into the future that we must set our plans. The master production schedule should be far enough out into the future that we can effect changes in our capacity rate, or changes in the material availability. The planning horizon has to be the greater of (1) the time needed to change capacity or (2) the cumulative lead time of our components. The cumulative lead time of the components is the sum of the longest planned lead times for each level, progressing down the entire bill of material. This can be seen in Figure 6–8, which shows that the purchase of the tubing is the *critical path;* that is, the longest "stacked" time for material availability to build the pen.

Because the planning horizon is a guideline, it is better to state it too long rather than too short, but it must be at least the longer of these two major factors. This point is so important that we should define it as the planning time fence. Our horizon needs to be long enough to let us feel comfortable with our ability to produce all the products.

The planning time fence for an individual part is also very significant for customer order promising. The best way to understand this is to look at an example. You are the master scheduler for pens at FineLine Industries. Our best customer calls in and asks us if we can ship 300 of Pen 42-208 in Period 8. We can quickly determine, looking at Figure 6–9, that we don't have sufficient master schedule orders planned to fill this demand. You will have to answer two questions. First, "Will it fit in the production plan, and can we schedule one large order in

FIGURE 6–8
Critical Path

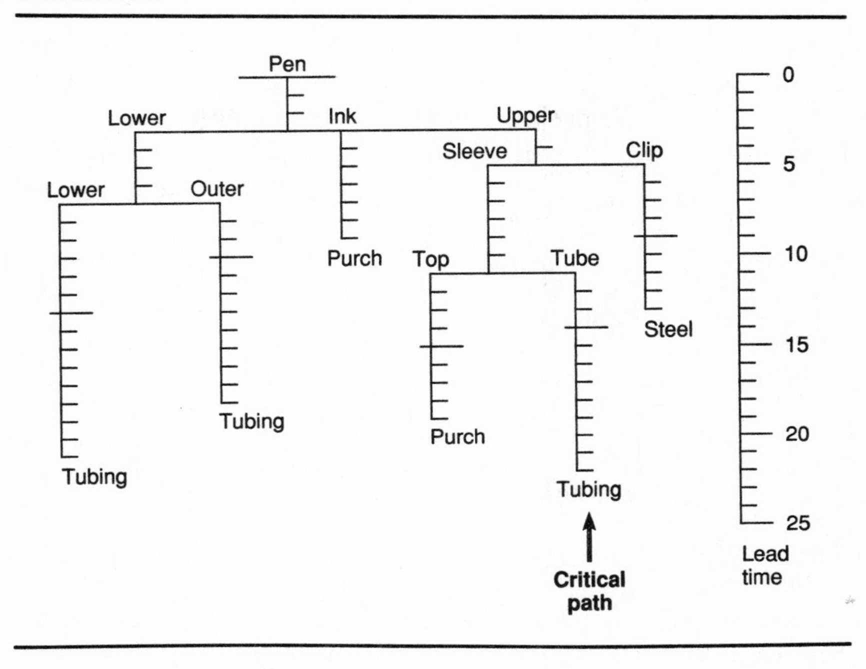

FIGURE 6–9
Master Production Schedule (demand and planning time fences)

Pen 42-208	Now	Period 1 4/23	2 4/30	3 5/7	4 5/14	5 5/21	6 5/28	7 6/3	8 6/10
Forecast		10	10	10	10	10	10	10	10
Actual orders		8	4	2		2			300
On hand	30	20	10	0	20	10	0	30	20
Available-to-promise		16	16	16	44	44	44	84	84
Master schedule					30			40	300

↑ Demand time fence ← Trade-off zone → ↑ Planning time fence

Period 8?" If our production plan is 300 per period, we can only put this new order in if we can move all the other pen orders into different time periods. This process is making the master schedule consistent with the production plan. Your second question is, "Do we have sufficient material to build the 300 pens?" If our planning time fence is at Period 7, we know that we can order new raw materials and have them arrive in time to build the production order for 300 pens. Hence, customer requirements beyond the planning time fence can be easily incorporated into the master schedule.

There is a powerful concept known as *what-if capability,* which implies that we should be able to simulate our production schedule to determine whether we can put the order for 300 pens into Period 8. We run the master schedule against the rough-cut capacity plan to see what other actions we may have to take in order to fill this new customer requirement. We could decide to split the order into two parts to smooth the shop load, for example, and now would be the time to discuss this alternative delivery plan with the customer. All of this serves the purpose of making sure that the master schedule is realistic before we go through the gross-to-net MRP logic.

Returning to the subject of time fences, let's consider another possibility. Suppose our customer wants delivery of the order of 300 pens in Period 1. If our lead time shows that it takes one period to build the pens in our shop, we would have to say that even if we have all of the materials on hand, it would be catastrophic to our shop to stop all the orders currently in production in order to build this new customer requirement. The *demand time fence,* then, is defined as the manufacturing lead time to build the product. Some companies choose to freeze the master schedule at the demand time fence.

While we don't want to imply that we would *never* put a customer demand in front of the demand fence, we must be clearly aware of the fact that we might throw the shop into chaos by doing so, whereas any reasonable request from the customer beyond the planning time fence can be implemented with minimum disruption to the shop. These time fences, therefore, are guidelines alerting us to the possible severity of our actions. From a practical viewpoint, we will accept almost any change beyond the planning time fence and resist nearly any change inside the demand fence, while the area between the two could be called the *trade-off zone*. If the customer has a request, it should be reviewed by the master scheduler to see if materials can be traded from one order to another to enable us to fill the customer request.

Continuing with our example, the customer wants 300 pens with black ink cartridges, but we do not have such a large quantity master scheduled. Looking at our records, you see that we *have* scheduled blue ink cartridges in that quantity, and therefore, in any case, we have sufficient tubing scheduled to come in to make upper and lower barrels for either color of pen. Since the blue cartridges have not yet been allocated to a customer order, we may have enough time to talk to our supplier and obtain black ink cartridges instead of blue, enabling us to build the customer request. This is an option because our commitment to the ink cartridge color does not occur until fairly late in the cumulative lead time for the components, as shown in Figure 6–8. We can call this a trade-off.

TIME-PHASED ORDER POINTS

These master scheduling techniques utilizing the stabilization effects of time fences are most applicable to dynamic product lines. If, on the other hand, our demand is smooth, uniform, and continuous, we can use a much simpler technique known as the *time-phased order point*. We use the fundamental MRP logic and apply it to independent demand items whose gross requirements come from a forecast.

As you can see in Figure 6–10, the gross requirements show 20 for every period. The MRP logic is applied against the gross requirements to create a projected balance on hand. This is

FIGURE 6–10
Time-Phased Order Point (independent demand)

	Now	1	2	3	4	5	6	7	8	9
		Period								
Requirement		20	20	20	20	20	20	20	20	20
Scheduled receipts				100						
On hand	80	60	40	120	100	80	60	40	120	100
Planned order releases					100					

Demand = 20 per period
Lead time = 4
Lot size = 100

On-hand balance = 80
Safety stock = 30
Scheduled = 100 due in 3

sufficient to generate planned order releases. These planned order releases would then become the input to drive the MRP system in planning the remaining dependent demand items.

This technique is best used for distribution center inventories or for service parts demands because it minimizes the amount of time spent by people in reviewing these planned order releases. While this is an advantage, at the same time we should guard against people's tendency to let the computer manipulate the master schedule for them.

THE MASTER SCHEDULER

Success in master scheduling depends a great deal on the capabilities of the master scheduler. This should be a person who is able to balance the demand for the product against the supply. The demand is the forecast and the sales orders, so the master scheduler should have good product knowledge. At the same time, he should also have good plant knowledge, to understand the capacity required and the flow of component materials.

Since the master scheduler's job is establishing the replenishment orders of the master scheduled items, he must understand that these orders drive the MRP system. The most effective actions that the MRP logic is going to recommend will be the planned order releases for the materials at the lowest level of the bill of material. So, as the master scheduled orders cross the planning time fence, we begin to make purchase commitments for the materials required. We can be most effective in controlling the MRP system when we focus on the point where the master production scheduled orders cross our planning time fence. (See Focus 6–B.)

Some companies freeze the master schedule at this point in order to obtain stability in the manufacturing process. If the lead time is longer than the customer's wait time, we have an impossible situation. Our lead time has become too long to effectively freeze the schedule and tell customers, "No." On the other hand, there is a point when forcing new orders into the shop produces dire consequences. If we can't freeze some portion of our master schedule, our lead time is too long, and we must reduce it to the point where we can freeze without penalty. It is impractical, after all, to talk customers into extending their time, and so the concept of excellent customer service requires that the change must come from our end.

The master scheduler is not out on this limb all by himself. This is one place where the management team should become involved in the process of creating the production plan and the parameters that control the master schedule. The production

FOCUS 6–B
Focus for Success

Focus at the point where the MPS crosses the planning time fence. This is the best spot to control MRP.

plan has to come before the master schedule, and the sum of the master scheduled items must equal the production plan. Therefore, it takes a concerted effort on the part of the management team to make the master production schedule realistic. This is particularly true if we allow the computer to automatically generate time-phased order points for a large part of our production. The forecast is marketing's commitment to what we will sell, the master production schedule is manufacturing's commitment to what we will make, the timing of these master schedules is materials' commitment to the availability of the components needed, and this becomes, ultimately, finance's commitment to what we are going to buy. This coming together is illustrated in Figure 6–11.

The best way to do this is the simplest. All the involved parties should come together in regular meetings to agree to commit our company to the master schedule, and then do whatever is necessary to make it happen.

FIGURE 6–11
Production Plan—Bring All the Elements Together

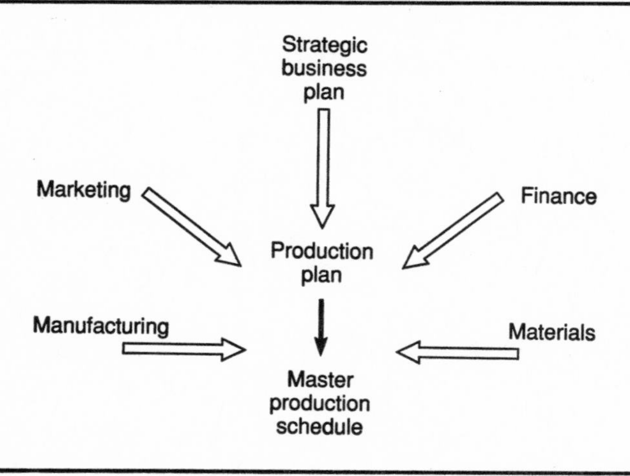

APPLICATIONS

All these master scheduling concepts can be used in various ways, depending on the specific end-unit products we wish to consider. In high-volume batch-process industries, where we are supplying, perhaps through a distribution system, we can use a distribution resource planning (DRP) system that makes the plant forecast the sum of the planned orders for the various distribution warehouses. There would be no customer orders per se, and the master schedule parts would feed directly into a distribution pipeline. The inventory service policy of the distribution system is a major input to determine the inventory levels. Process industries, therefore, take the production plan and the inventory policies to create the master production schedule, actually creating a build line without necessarily creating a time-phased format for every end unit. These build lines, nonetheless, are the schedules that drive the component MRP plans.

Another issue that we see in today's world-class companies is the selection of exactly which items are going to be master scheduled. More and more, companies are migrating away from master scheduling every discrete configuration. In the pharmaceutical world, we might master schedule aspirins by multiple thousands of tablets without regard to packaging size requirements. We would use a planning bill to establish estimated quantities of various package options—50-count, 100-count, 250-count, tins of one dozen. We would supply the main component in bulk, and package it at the last minute in various customer-specified sizes. This is similar to the distribution concept known as *postponement*, which says that we will not commit the product to exact customer configurations until the latest possible point in time. This gives us the same result as reducing the time fences—that is, the lead times—down to the smallest length possible. Companies that paint their products in assorted colors will not paint them or add the identification labels until the customer order is actually received.

Dealing with the concept of rough-cut capacity, another point is that our product load profiles do not have to be restricted to manufacturing labor. In high-volume process industries, the storage cube of a product may become the limiting factor, and we can use rough-cut capacity planning to calculate storage requirements. We also see airfreight delivery companies using the cube as a parameter on their schedules. In still other applications, it is quite common for companies producing high-value items to include dollar flow as part of the product load profile.

Department of Defense contractors find that MRP is most applicable for products that are repetitive, when the master schedule can be established with the purchasing agency. Then the bill of material can be stabilized and the anticipated build schedule set, so that we can use the master schedule and MRP logic to schedule the components and plan the capacity.

SECTION 3

OUTPUTS

CHAPTER 7

PLANNED ORDER RELEASES

Now we turn to the first of the three main outputs of our MRP system. Planned orders are the bottom line of our MRP time-phased chart. These planned orders are created when there is a net requirement and the computer system applies lot sizing and lead time offset to arrive at a planned order release date. These planned orders are moved around both in schedules and quantities by the computer system, using the basic logic of MRP. We expect that those suggestions may be changed, deleted, or increased as conditions change.

When the planned order release date for one of these suggestions appears in Period 1, this notifies the planner that it is time to take whatever action is necessary to start that order through the purchase or the manufacturing process. What we really want to do is to convert our planned order release into a scheduled receipt. This will show up on the MRP worksheet as an *exception message*. In Figure 7–1, an exception message has been printed that suggests to our planner that she should release an order to build 25 upper barrels, starting in Period 1 and finishing in Period 4. This is a continuation of the basic MRP logic that we saw in Figure 3–11.

Of course, this information can be seen on the MRP worksheet for the individual item, but it is also effective to have the computer system print out a listing of all the parts that have these types of suggestions. This is known as an *exception report*. It can be on the planner's desk at the beginning of the workday, listing all the parts to be reviewed during the day, because the suggestions apply to things that should take place in Period 1. The purpose of the suggestion is for the planner to convert the

FIGURE 7–1
MRP Worksheet (upper barrel)

Lead time = 3

	Now	1	2	3	4	5	6	7	8	9
Requirement		20		20	20		5		20	
Scheduled receipts			25							
On hand	25	5	30	10	15	15	10	10	15	15
Planned order releases		25				25				

Suggest new order: 25 due Period 4 Rel: Period 1
25 due Period 8 Rel: Period 5

planned order release into a scheduled receipt. This takes the form of a *manufacturing shop order*, which should be printed and released to the manufacturing shop floor. If it is a purchased part, we should release a requisition to purchasing, who in turn will release a purchase order to our vendor to authorize it to start production. When these steps are done, we would also update our dynamic inventory data record, showing that a shop work order has been released or a purchase order generated. Once these records are created, the next generation of the MRP report will show these as scheduled receipts, as shown for the upper barrel in Figure 7–2.

As you can see, the requirements have "rolled ahead" one period, and our balance on hand has been reduced by the 20-piece requirement that was in Period 1. The balance on hand is now down to 5, but we show both scheduled receipts: the one that was there last week and the one we just released. The normal MRP logic goes on to evaluate the new requirements. We are viewing this process in what is called the *regeneration* method of MRP. Simply put, this means that we regenerate the

FIGURE 7–2
MRP Worksheet (upper barrel—next week)

Lead time = 3

		Period								
	Now	1	2	3	4	5	6	7	8	9
Requirement			20	20		5		20		10
Scheduled receipts		25		25						
On hand	5	20	10	15	15	10	10	15	15	5
Planned order releases					25					

Suggest new order: 25 due period 7 Rel: Period 4

whole MRP logic in each time period, most commonly weeks. We can see this very clearly when we compare Figure 7–1 to Figure 7–2.

RELEASING MANUFACTURING PARTS

It is important to note that the planned order releases are controlled by the computer system, but the scheduled receipts are controlled by people. Once a scheduled receipt is established, it will not be changed by the computer system because people have already made commitments for productive capacity, lower-level components, or vendor agreements.

It is critical at this point that we commit our resources to this replenishment order. The commitment is for both the order quantity and the scheduled due date. First, we will examine the case of manufactured parts. We have our same two problems, material and capacity availability. We must first check the component availability. We examine the bill of material to determine those parts that must be checked, to see if we have sufficient balance on hand of each of the components to complete our order. This can often be done on-line, and the computer program will highlight shortages only. As we review the component availabilities, we should also evaluate the quantity for efficient raw material utilization and other order quantity modifiers that we discussed when we examined the inventory data file. This is also the time for us to evaluate any other factors that might affect our situation—such as how many parts fit into a container—and influence the effective flow of the order through the shop.

At the same time, we should examine the capacity required to build our suggested order quantity. It is time now to evaluate setup operations, and if any capacity constraints exist in the manufacturing flow, we may need to change the order quantity or the order date. These constraining factors include such things as machines undergoing scheduled maintenance, operators on vacation, or other problems that may show up on our capacity requirements report. It is also important to consider the flow of the products through the shop. Downstream machines might have a large backlog that could affect our decisions regarding the quantity or date of a specific order. It makes no sense to release the order if we cannot realistically expect to complete it by the due date. If we have a capacity or materials problem, we should deal with the problem now, rather than aggravate it further by releasing new orders. We should commit, on a deep, personal level, to our ability to make the part on that date.

Now we can tell our system that we released the planned order release for the specific part number, quantity, and due date. At this point, we expect to print a *manufacturing shop order*. Since all of this processing can take a considerable amount of time, we should allow time to do it prior to the release of the order to manufacturing. The planner should begin to look at the planned order releases far enough ahead of time to keep the paperwork from becoming part of the item lead time.

When the planner decides that it is time to physically release the order, she will generate the manufacturing shop order.

This is actually a document like the one shown in Figure 7–3. It is the authorization to produce a specific quantity of a specific part number. Usually accompanying this authorization are the other documents necessary to show how the part is to be built. Called the *shop packet,* this typically includes engineering drawings, process sheets, and so on.

Although there are differences of opinion as to whether this authorization should include the original order due date, we think that the date should be left off because the main tool of priority control is the dispatch list, which we will discuss in detail in Chapter 10.

One of the shop packet documents is the picking list or

FIGURE 7–3
Manufacturing Order

Item: Part no. 42-208
Pen, gold, blue ink

Work order: 37105
Date printed:

Engineering drawing:
Date: Revision:

Start date:
Planner:

Order quantity: 52
Unit of measure: Each

Bill of Material

Component Part No:	*Description*	*Total Quantity*	*Unit of Measure*
20–231	Lower subassembly, gold	52	Each
20–201	Upper barrel, gold	52	Each
37–7213	Ink cartridge, blue	52	Each

Routing

Operation Number	*Work Center*	*Description*	*Tools*	*Setup Time*	*Run Time*
010	D15	Assemble complete		0.0	0.1
020	S10	Pack			

kitting list. This document lists the components that are to be pulled from stock, as shown in Figure 7–4. The stockroom employees will pull the quantities of components needed, gather them together, and deliver them to the first manufacturing operation. This is commonly known as *kitting* or *staging*. The order proceeds through the successive operations as described in the routing, which we discuss in Chapter 9.

The picking list is the documentation used to reduce the

FIGURE 7–4
Pick List

Item: Part no. 42-208
Pen, gold, blue ink

Work order: 37105
Date printed:

Engineering drawing:
Date: Revision:

Start date:
Planner:

Order quantity: 52
Unit of measure: Each

Bill of Material

Component Part No:	*Description*	*Quantity For One*	*Quantity Total*	*Unit of Measure*	*Pulled*
20–231	Lower subassembly, gold	1.000	52	Each	____
20–201	Upper barrel, gold	1.000	52	Each	____
37–7213	Ink cartridge, blue	1.000	52	Each	____

Date work order pulled: ____________________

By: ____________________

Update inventory: ____________________

balance on hand for the components as the parts are moved from stock and delivered to work in process. When the assembly is completed on the shop floor, this process is reversed. The assembly, in this example, Pens 42-208, will be added to our finished goods inventory. At that point, the manufacturing shop order is closed.

RELEASING PURCHASED PARTS

The process for a purchased part is very similar, for we must consider supplier capability and whether it is realistic to expect that the vendor can produce the quantity suggested by the due date. If the answer is yes, then we should release the requisition to purchasing, who will then contact the vendor, negotiate a purchase order, and release it to that vendor. Part of the negotiation should be the vendor's agreement to the purchase order due date, so that the scheduled receipt represents the vendor's commitment to meet that date.

Once again, this process can be very time-consuming, and to accurately represent the lead time for purchased parts, we must include the vendor's manufacturing time, plus the transportation, plus our receiving time, plus the time it takes to prepare the purchase order. The part of this time line that we have the most control over is the preparation of the purchase order paperwork. We must strive to have all order preparation time occur prior to the items' lead times. One of the more common approaches to this is the blanket order, in which the purchasing department has already negotiated with the vendor and we can send our release directly to the vendor. The mechanics of handling a blanket order varies among software systems, but the primary point to remember about blanket orders in this context is that they provide another way of keeping the lead time as short as we can make it.

SAFETY STOCK

Another issue that comes up when we consider planned order releases is the concept of using safety stock to take care of fluctuations in demand or supply. The premise is that if we have safety stock for these parts, we don't need to be quite as precise about our dates. For purchased parts, safety stock takes care of fluctuations in the rate of supply. Using this principle and applying it to lead time, what is implied is that the lead time might have variations; an effective method of compensating for these variations is *safety lead time.*

Safety lead time is the amount added to the supplier's lead time for the purpose of completing the receipt prior to the need date. There are two common ways of handling this concept. The less desirable approach is simply inflating the vendor's lead time. For example, in Figure 7–5, for our upper barrel tops, the vendor says that the lead time is two periods, so we add one week safety time and the lead time looks like three periods. A more appropriate way of doing this is to show both lead times on our inventory data, showing the vendor lead time as two

FIGURE 7–5
Safety Time: Inflating Vendor's Lead Time (upper barrel tops)

Vendor time = 2
Safety time = 1
Total time = 3

	Now	1	2	3	4	5	6	7	8	9
Requirement		10	20	10		15				
Scheduled receipts			25							
On hand	20	10	15	5	5	15	15	15		
Planned order releases			25							

periods and the safety time as one period. The planned order release in Period 2 is expected to come into stock in Period 4, which inflates our inventory in Period 4. This is demonstrated in Figure 7–6, which actually shows what we expect to happen. Really, however, the best way of handling this whole question is to have a completely reliable relationship between our company and our supplier.

Safety stocks are also used to accommodate fluctuations in demand. In an MRP system with dependent demand items, we have the assumption of certainty of demand, which means that once the master production schedule is established, those requirements are exploded to create the gross requirement for components. If we have a realistic master production schedule that we can keep stable by using time fences, we should not expect much fluctuation in the demand rate for components. The only place where we should encounter fluctuations in demand is at the master schedule level. It is appropriate, therefore, that we plan for safety stock only at the MPS level. Using safety stock in components only tends to disrupt the true priorities of component orders.

FIGURE 7–6
Lead Time plus Safety Time: (upper barrel tops)

Vendor time = 2 Safety time = 1	Now	Period 1	2	3	4	5	6	7	8	9
Requirement		10	20	10		15				
Scheduled receipts			25							
On hand	20	10	15	5	30	15				
Planned order releases			25		↑	↑				

Head time   Safety

As an example, let's suppose that our upper barrel tops are really required as shown in Figure 7–7. We can see that the planned order release in Period 3 is due to be received in Period 5. Now if we add a requirement for safety stock of 10, as shown in Figure 7–8, this causes the projected balance on hand to drop below the safety stock in Period 3, initiating a planned order receipt in Period 3 that must be released in Period 1. Therefore, the introduction of safety stock not only increases the projected balance on hand, it causes us to receive an order in a time period before we actually need it. In actual practice, having safety stock causes us to generate an expedited order in advance of the true need date. Masking the true need date causes a lot more complications when we consider multiple levels and differing gross requirements. We create a series of situations that can ultimately yield skewed results.

In summary, remember that if we must have safety stocks, they should only be at the master production schedule level to accommodate fluctuations in customer demands. For fluctuations in the rate of supply, safety time is an appropriate alternative. Safety stock should never be introduced at the interme-

FIGURE 7–7
MRP Worksheet (upper barrel tops)

Lead time = 2 Lot size = 25	Now	1	2	3	4	5	6	7	8	9
Requirement		10	20	10		15				
Scheduled receipts			25							
On hand	20	10	15	5	5	15				
Planned order releases				25						

(Period header spans columns 1–9.)

FIGURE 7–8
Safety Stock (upper barrel tops)

Safety stock = 10
Lead time = 2
Lot size = 25

	Now	1	2	3	4	5	6	7	8	9
Requirement		10	20	10		15				
Scheduled receipts			25							
On hand	20	10	15	30	30	15				
Planned order releases		25								

Period

diate levels of the bill of material. The overriding factor in controlling planned order releases is to manage the lead times so that they are as short and as dependable as we can possibly make them, as shown in Focus 7–A.

FOCUS 7–A
Focus for Success

Manage *lead time* to be as short as possible.

APPLICATIONS

There is an implied assumption in this discussion of ensuring the availability of components before we release manufacturing shop orders. All of the components must be in stock in order to build an assembly. However, many companies today are able to flatten their bills of material by, in effect, releasing the orders to manufacture components and the order to do the assembly at the same time. This is done when we use the flow manufacturing process rather than the batch process that the MRP normally assumes. These component part orders are usually marked or flagged in some way to indicate that they are to be used directly on the shop floor at the next higher level of assembly. This concept can be done using the phantom bill of material technique, or with other techniques such as special coding of make-to-order components or visual signals. To implement these techniques, planners will often earmark certain orders by showing that they are required at higher levels. Typically, they will also mark the assembly order if the components are being manufactured and are due on a certain date.

If we have a high-volume, repetitively manufactured part that is supplied to customers, we can do away with the master scheduling concept and move toward a demand-pull process. What this means is that we only use master production scheduling as the signal to the suppliers of raw material requirements, without actually releasing the orders for manufacturing. Rather, we can use customer orders, so that whatever was shipped from finished goods inventory yesterday will determine the quantity to be released for manufacturing today. For example, there is a company that uses this concept very effectively with orthopedic medical products. Its products are converted from one raw material into a finished good. The company uses the master production scheduling process to plan the orders of raw materials, and it only releases manufacturing orders to replace finished goods that have been sold.

CHAPTER 8

RESCHEDULE NOTICES

The second major output from the MRP system is the *reschedule notice.* The act of converting planned order releases into scheduled receipts establishes an order quantity and a due date. That date will be valid because of the basic logic of MRP at the time of the planned order release. However, there's an old factory saying that whatever the conditions are at 7:00 A.M., they're likely to change by 7:15. What this really means is that the manufacturing environment is very dynamic and in a constant state of change.

The main purpose of an MRP system is to indicate the actions we need to take regarding the flow of materials, both purchased and manufactured, whenever conditions change. We need to emphasize right at the outset that if the planned lead time (the time between the release date and the due date) is extremely long, the likelihood of having to make a change at some point during that lead time period is very high. It is in our best interests, therefore, to keep the lead times as short as possible in order to minimize the number of changes we will need to make, as shown in Focus 8–A.

Remember the basic definition of an MRP system: A scheduling technique that has as one of its main objectives keeping the due date equal to the need date. This means that when an order is a scheduled receipt, it has a planned due date. But as our requirements change, the need date could conceivably change, as well. As an example, see Figure 8–1, where our upper barrel is shown with a scheduled receipt due in Period 2. If we examine this MRP worksheet closely, we can see that the projected balance on hand will go up to 30 during Period 2 and

FOCUS 8–A
Focus for Success

Focus on reducing *lead times* in order to minimize changes.

that the scheduled receipt is not actually needed at that time. We can fill all of the gross requirements with the current balance on hand until the requirement in Period 3. Therefore, the MRP system would print out an exception message suggesting that we reschedule the receipt for 25 pieces in Period 2 out to Period 3.

It is important to note that the computer system will not actually change the due date of the order because actions have already been taken by people to begin building the part. So this is merely a suggestion that *a person* should evaluate whether or not it is economically sound to change the order. There may be a number of reasons why delaying the order is not a smart idea.

FIGURE 8–1
MRP Worksheet (upper barrel)

		Period								
	Now	1	2	3	4	5	6	7	8	9
Requirement		20		20	20		5		20	
Scheduled receipts			25							
On hand	25	5	30	10	15	15	10	10	15	15
Planned order releases		25				25				

Suggest reschedule: 25 from Period 2 to due Period 3

The work in process has already been committed, for example, or downstream machines need the work, and perhaps we should go ahead and accept the higher inventory and bring the order in early.

On the other hand, if we wish to maintain minimum inventory investment, it might be smart to delay delivery of this order. The purpose of the MRP system is to make suggestions to take these sorts of actions. This is similar to the exception notices to release a planned order, in that human judgment should be exercised before action is taken. These reschedule notices for all parts could be brought together in an exception report that would list all the suggested changes throughout our operation.

If the item in question is a purchased part, then our decision involves another company. We need to coordinate with the supplier's production control department because we suspect that they have an order going through their process to build our part. Many companies find that this is the spot where it is beneficial to have their planners talk directly to the vendor's planners. (Sometimes the position is called the buyer/planner.) The traditional purchasing concept says that the buyer should set up the original contract with the vendor. Sometimes these contracts will be one-shot deals, sometimes they will represent long-term commitment, as in the case of a blanket order. In any case, the job of maintaining the scheduled dates on each scheduled receipt is our planner's responsibility as she talks with the vendor's planner.

EXPEDITING

In the context of a book, we can deal with the example of an order being delayed, but most people will point out that their own experience comes from orders that are expedited. Expediters are usually known as the "hurry up" group, and they end up trying to get the order in earlier than it was planned because the requirements have increased rather than been deferred.

When this is the case, we don't have the luxurious option of disregarding the suggestion from our MRP system without looking at the higher levels of the bill of material. Here, we would use pegging to tell us which upper-level assemblies created the gross requirement to identify the assemblies that would be delayed, all the way up to the master schedule, if necessary. Now we are dealing with potential customer delays that might adversely affect our business. Our first emphasis, as George W. Plossl, CFPIM, advocates in his presentations and seminars throughout the business community, must be to "move heaven and earth" to make the schedules.

But what caused the gross requirements to change? Logically, the reason why our gross requirements change is that a change was made in a higher level of the bill of material. There are all sorts of reasons why this could happen. The lot-sizing rules may have changed, the bill of material might have changed, the inventory balances could have been inaccurate, or a host of other things, any of which would cause the planned order releases to change. All of these problems stem from data integrity problems, and we should be very careful to track the cause of the discrepancies and correct the causes, not just the errors, as we discussed when we examined inventory data.

DYNAMICS OF MRP

However, the change that must be dealt with most frequently is a change in the master schedule, which is most often precipitated by a change in a customer's demand. It is the purpose of the MRP to show us what actions we need to take to satisfy this new customer demand. This can best be understood through an example.

You are the planner for pencil components at FineLine Industries. In the MRP worksheet for erasers, Figure 8–2, we have a scheduled receipt for 600 pieces due in Period 4, and a planned order release for 800 pieces that should be released now

FIGURE 8–2
MRP Worksheet (eraser assembly)

Lead time = 5

	Now	1 4/23	2 4/30	3 5/7	4 5/14	5 5/21	6 5/28	7 6/3	8 6/10
Requirement			1,600		1,000		800		
Scheduled receipts					600				
On hand	2,000	2,000	400	400	0	0	0	0	0
Planned order releases		800							

Period (header spanning periods 1–8)

Suggestion: Scheduled receipt: Order 39042 for 600 due 5/14
New order for 800 due 5/28 Release now

for delivery in Period 6. When you look at the next regeneration of the MRP, in Figure 8–3, you should make several observations. First, the gross requirements have changed. You can see that the quantities have rolled ahead one period, with the exception of the 1,600 pieces that have been pushed out to Period 6. Also, you did not release the planned order for 800 because you knew of the impending customer change in requirements. So, you change the master schedule and rerun the MRP to let the suggestions show you what actions need to be taken.

In Figure 8–3, there are two suggested actions. One is a reschedule notice to move the scheduled receipt of 600 pieces from Period 3 to Period 6. The other notice is a suggestion to release a different planned order for 800, due also in Period 6. After you take that action, the next MRP worksheet would look like Figure 8–4, which indicates that you have taken the action necessary to fill the customer demand while at the same time minimizing inventory investment.

FIGURE 8–3
MRP Worksheet (eraser assembly—next week suggestions)

Next week

Lead time = 5

	Now	1 4/30	2 5/7	3 5/14	4 5/21	5 5/28	6 6/3	7 6/10	8 6/17
Requirement				1,000		800	1,600		
Scheduled receipts				600					
On hand	2,000	2,000	2,000	1,600	1,600	800	0	0	0
Planned order releases		800							

Suggestion: Reschedule receipt: Order 39042 for 600 due 5/14 to 6/3
New order for 800 due 6/3 Release now

FIGURE 8–4
MRP Worksheet (eraser assembly—actions taken)

Next week

Lead time = 5

	Now	1 5/7	2 5/14	3 5/21	4 5/28	5 6/3	6 6/10	7 6/17	8 6/24
Requirement			1,000		800	1,600			
Scheduled receipts						600 800			
On hand	2,000	2,000	1,000	1,000	200	0	0	0	0
Planned order releases									

Schedule receipt: Order 39042 for 600 due 6/3
Schedule receipt: Order 39043 for 800 due 6/3

This is the real function of an MRP—to give us suggestions of the actions we should take to accommodate customer demands at the MPS level. Of course, in our example, we have shown a requirement delayed, pushed out in time, causing us to push orders out in time as well. In the real world, the push is frequently in the other direction. As customers change their demands and request accelerated delivery, we see our requirements being shortened in time, causing us to expedite component delivery. In even the simplest of manufacturing companies, there are many master schedule items with many component parts where this dynamic of change is going on daily. There can be a real ebb and flow of change at the master production schedule, causing us to be in a constant state of fluctuation. Remember the other changes we mentioned earlier, which may have stemmed from cycle count adjustments, unplanned usages, or orders closed short. All of these factors can combine to cause what is sometimes called the *nervousness* of an MRP system. The average person working with these reports and exception messages may tend to be frustrated at the futility of trying to maintain the dates.

FIRM PLANNED ORDERS

One technique that is used to reduce the nervousness is called the *firm planned order*. A firm planned order is a planned order release that is flagged to fix it both in quantity and in time. If you recall the basic logic of MRP, a planned order release is normally moved around by the computer system. The planned order releases at one level create the gross requirements at the next level, so moving a planned order release will rearrange the gross requirement, precipitating lower-level nervous rescheduling. We can minimize this by not allowing the computer to change our planned order releases. As the planner responds to material or capacity problems, at some point it becomes advisable to go ahead and implement her solution. We want to tell the computer system that we have made plans and we intend to implement them, which we do by selecting orders and firming up the planned order releases.

To describe this process, let's take an example. In Figure 8–5, we see that the upper barrel has a planned order release for

FIGURE 8–5
MRP Worksheet (first iteration)

Upper barrel

Lead time = 5
Lot = 3 periods

	Now	1 5/14	2 5/21	3 5/28	4 6/3	5 6/10	6 6/17	7 6/24	8 7/1	9 7/8	10 7/15
Requirement		30	20	25	25	25	75	20	40	30	40
Scheduled receipts											
On hand	250	220	200	175	150	125	50	30	70	40	0
Planned order releases				80							

Clips

Lead time = 2
Lot = Lot-4-lot

	Now	1 5/14	2 5/21	3 5/28	4 6/3	5 6/10	6 6/17	7 6/24	8 7/1	9 7/8	10 7/15
Requirement				80							
Scheduled receipts											
On hand	0	0	0	0	0	0	0	0	0	0	0
Planned order releases		80									

New order for 80 due 5/28 Release now

80 pieces in Period 3, due in Period 8. We also know that planned order releases in one level create the gross requirement in the next, so the order for 80 upper barrels creates a gross requirement for 80 clips in Period 3. Notice that the balance on hand for clips is 0 and the lead time is 2, so we have a planned order release for clips in Period 1, due in Period 3.

Now let's roll the MRP ahead one period, and notice a change in requirements for the upper barrel in Figure 8–6, probably caused by changes in customer orders. This change in requirements has caused a change in the planned order release to show that 95 should be released in Period 3 for delivery in Period 8. Once again, the planned order release for 95 upper barrels creates a gross requirement for 95 clips needed in Period 3, but from our last run of the MRP, we have released an order for 80 clips, so it now shows up as a scheduled receipt for 80 due in Period 2. The MRP logic for the clips would suggest, therefore, a reschedule notice to move our recent order out to Period 3 and at the same time release a brand-new order for 15 more. This can drive the people who receive such orders to the brink of insanity; obviously, management doesn't know anything about what they're doing. To break the cycle, let us move on to Figure 8–7.

Here, we see the requirements for upper barrels as originally shown in Figure 8–6, with the planned order release for 80 pieces to be released in Period 2 and due in Period 7. This was the original suggestion made in Figure 8–5, and that order has been flagged as a firm planned order (FPO). Now the computer system will not change either the quantity or the date, and it still lists this order as starting in Period 2. Remember that we have rolled the requirements ahead by one period, so the real calendar date for the planned order release has not changed. However, we can tell by looking at the projected balance on hand that this planned order release could be delayed one period, and so we get a suggested reschedule of that firm planned order. This is because the MRP logic will not alter the quantity or the date of a firm planned order, even though it is on the planned order release line.

FIGURE 8–6
MRP Worksheet (changed requirements)

Upper barrel

Lead time = 5

Lot = 3 periods

	Now	1 5/21	2 5/28	3 6/3	4 6/10	5 6/17	6 6/24	7 7/1	8 7/8	9 7/15	10 7/22
Requirement		20	25	20	25	75	20	40	50	40	30
Scheduled receipts											
On hand	250	230	205	185	160	85	65	25	70	30	0
Planned order releases				95							

Clips

Lead time = 2

Lot = Lot-4-lot

	Now	1 5/21	2 5/28	3 6/3	4 6/10	5 6/17	6 6/24	7 7/1	8 7/8	9 7/15	10 7/22
Requirement				95							
Scheduled receipts			80								
On hand	0	0	80	0	0	0	0	0	0	0	0
Planned order releases		15									

Reschedule receipt: Order 40223 for 80 due 5/28 to 6/3
New order for 15 due 6/3 Release now

Continuing the logic of MRP for the upper barrels, we see the suggestion for a second planned order release for 15 more pieces starting in Period 5 and due in Period 10. Those planned order releases cause the gross requirements for clips to show 80 required in Period 2 with 15 more required in Period 5. This matches up with our scheduled receipt, the order for 80 that we started last week. The MRP suggestion is to continue to build

FIGURE 8–7
MRP Worksheet (firm planned order)

Upper barrel

Lead time = 5

Lot = 3 periods

	Now	1 5/21	2 5/28	3 6/3	4 6/10	5 6/17	6 6/24	7 7/1	8 7/8	9 7/15	10 7/22
Requirement		20	25	20	25	75	20	40	50	40	30
Scheduled receipts											
On hand	250	230	205	185	160	85	65	105	55	15	0
Planned order releases			80 FPO			15					

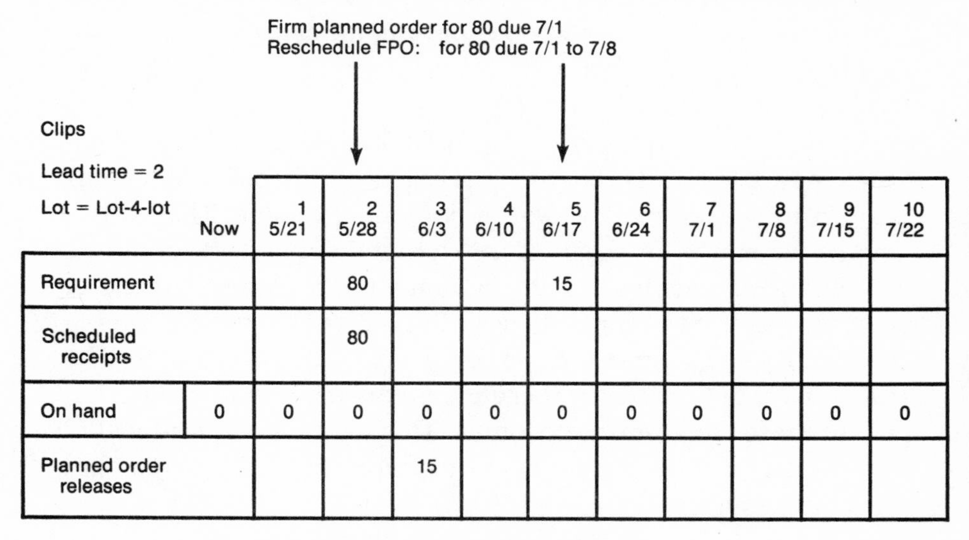

Firm planned order for 80 due 7/1
Reschedule FPO: for 80 due 7/1 to 7/8

Clips

Lead time = 2

Lot = Lot-4-lot

	Now	1 5/21	2 5/28	3 6/3	4 6/10	5 6/17	6 6/24	7 7/1	8 7/8	9 7/15	10 7/22
Requirement			80			15					
Scheduled receipts			80								
On hand	0	0	0	0	0	0	0	0	0	0	0
Planned order releases				15							

Receipt: Order 40223 for 80 due 5/28
New order for 15 due 6/17 Release 6/3

that order as planned but to plan to release a new order for 15 more, starting in Period 3 and due in Period 5. We will not, however, convert that planned order release into a scheduled receipt until three more periods have passed, and in that interval, we would suspect that the requirements for upper barrels will change several times. We have stabilized our manufacturing process to go ahead and build the 80 clips due in Period 2, and we expect that the next order will come along in about Period 5. There are two or three more periods before we actually have to commit to what that quantity will be.

The beauty of the firm planned order is that as we firm up planned order releases at the higher level, we impose stability on our manufacturing processes at lower levels. This stability is appreciated by our own manufacturing department and by the suppliers of purchased parts. The ultimate use of the firm planned order is at the master production schedule, which will tend to stabilize all of the lower levels. The objective is to firm the orders for master scheduled parts as those planned orders cross the planning time fence, as we discussed in Chapter 6. If we really want to stabilize our manufacturing shop, we should focus on stabilizing the master production schedule, as shown in Focus 8–B.

Here again, we encounter an illustration of the vital need for keeping the lead times as short as possible in order to use our firm planned orders. We have a potential problem: The MRP is supposed to give us suggestions so that the due dates of orders match the need dates. As we implement firm planned orders or convert planned orders into scheduled receipts, we are forcing the MRP logic to make *suggestions only* for those changes. The farther out in time we fix these plans, the more voluminous the list of these suggestions becomes. This makes it very unwieldy to operate, and the users begin to think that the system isn't flexible enough.

FOCUS 8–B
Focus for Success

Focus on firming up the MPS in order to stabilize manufacturing.

There is a way to have flexibility and stability at the same time, and that is to fix these order quantities and dates on the shortest lead time possible.

Remember the planned lead time listed in our inventory item data is exactly that—*planned* lead time. As we receive suggestions to move due dates in and out to match our need dates, this may cause the actual lead time to differ from the planned lead time. This is a very effective way to meet changing customer requirements, once we understand that lead time can be flexed to accommodate customer demands. We will examine this idea in detail later, when we move on to the subjects of managing MRP priorities and managing capacities.

APPLICATIONS

All through this chapter, we have talked about shortening lead times to gain maximum flexibility. We see many world-class companies doing this with their purchased parts by setting up blanket order contracts that establish commitments between themselves and the suppliers for capacity rather than for specific parts. The planners of both companies are then allowed to do the vendor scheduling. This takes teamwork with the vendors, a concept that usually is designated *supplier certification*. One company we know of calls this *schedule sharing*. These concepts imply quality and schedule reliability. We know of several companies who have established delivery patterns, such as steel deliveries every morning. The planner can call the vendor as late as 4:00 P.M. to order materials that will be on the next morning's truck. A long-term commitment has been established with the vendor to define which parts can be ordered with this quick response. This allows the planners to convert planned order releases into scheduled receipts in a one-day lead time. We see high-volume process industries getting these rapid-response deliveries for their packaging materials; and companies ordering maintenance and repair order (MRO) parts and fasteners from distribution centers receive those items in this quick response mode.

We know of one high-speed company that can call vendors at 3:00 A.M. and get delivery before dawn. You have to have a close long-term relationship with your suppliers to be able to do this kind of thing, but the results in enhanced customer service can be spectacular.

Several companies have been innovative with the concept of "milk runs," where trucks may pick up from several vendors and deliver to several customers on a routine basis. This takes the ultimate in cooperation among multiple vendors and multiple customers, but again, the payback in enhanced shipping efficiency is tremendous.

CHAPTER 9

CAPACITY REQUIREMENTS PLANNING

In our previous two chapters, we discussed the two main outputs of the MRP logic, namely, planned order releases and reschedule notices. The actual pieces of data maintained by the system are part number, quantity, and due date. This data is actually tracked for the planned order releases and that special case, firm planned orders, where we flag an order so that the system cannot change the date and quantity. Of course, the scheduled receipts represent those orders that have already been released with actual quantities and actual order due dates.

ROUTING

We are now going to examine manufacturing orders. For any part that we manufacture ourselves, we must have a routing that describes how the part is produced. The routing is defined as the set of information describing the sequence of steps required to manufacture the product. Each of these steps details the operation to be performed, the location, which can be a machine or a work center, and the setup and run-time standards for the part. Many companies include other information about how the operation is performed on the product, such as the tooling required, operator skills, crew size, and testing requirements.

A work center is a grouping of one or more machines and/or people that are able to complete a specific function or process. Historically, this once meant similar types of machines, but today, it can be expanded to include a work cell or a production line that produces a full product. The purpose of defining these work centers is to permit us to consider them as one unit when scheduling, planning capacity, and measuring performance. We keep a list of these work centers in the system, typically in what is called a *work center file.*

An example of a routing for the upper barrel of our pen is shown in Figure 9–1. Once we have established our routing, we can take our MRP output, the order quantity and due date, and calculate the total number of hours required to produce the part on each work center. We know the date that the order is due, and the date when the order should be started. Accordingly, we can estimate when each operation needs to be performed so that we can complete the order as scheduled. We must consider setup and run times for each operation, plus one additional factor known as *queue time.*

A queue is, simply put, a waiting line. Queue time, in a manufacturing sense, is the amount of time an order waits in front of a work center before the people in that work center start to work on that job. For example, in our upper barrel, the first operation is to cut the tubing. The second operation is to taper the end of the tubing. If we look at the operation times, we

FIGURE 9–1
Routing (upper barrel)

Item Number	*Description*	*Unit of Measure*	*Engineering Revision*	*Drawing Number*
20–201	Upper barrel	Each	D	1391

Operation Number	*Work Center*	*Operation Description*	*Tools*	*Setup Time*	*Run Time*
010	A30	Cut tubing	A–63782	1.5	.1
020	A20	Taper end		2.3	.14
030	C10	Press on top	Fx179	.5	.04
040	A50	Assemble clip	QC301	.5	.12

know that by taking the run-time standard, usually expressed as hours per piece, times the pieces, plus the setup time, we arrive at the total operation time.

The cutting of the tubing has a run standard of .1 hours per piece, times 30 pieces, representing 3.0 hours worth of work, plus 1.5 hours of setup, for a total of 4.5 hours operation time. Once the operation is completed, we know that the job is not instantly started through the next operation, which tapers the ends. If we consider the movement of material through our shop, there is a time when the cutting operator puts the material down and waits for a forklift driver to pick up the skid. When the material is moved to the taper work center, chances are that another order is being worked on at that work center already. So our order is placed in the queue, waiting its turn at the work center. These elements can be added together and called *inter-operation time*, or the time between the completion of one operation and the start of the next. We find in most plants that the largest element of that time is the queue time. The sum of the operation time plus the interoperation time is the amount of time we should actually allow for each operation, as shown in Figure 9–2. Notice also, that the lead time should include the time necessary to pull the raw material from stock.

FIGURE 9–2
Lead Time (upper barrel)

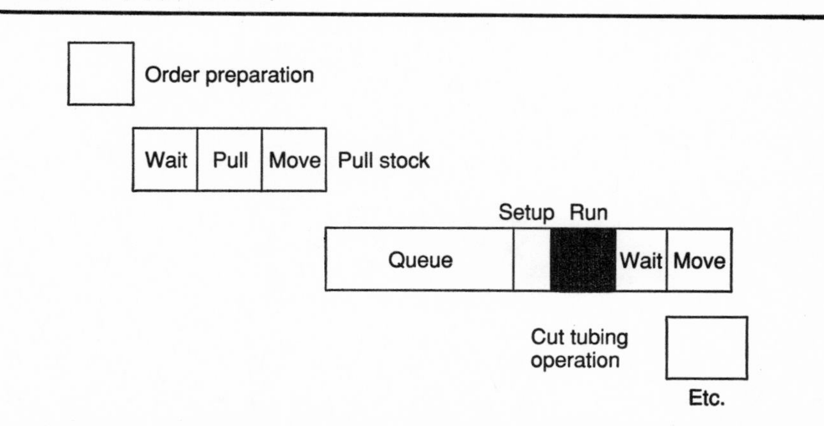

We can calculate the operation time from the order quantities, but the time allowed for interoperation time must come from estimates that we enter and maintain in the work center file. These numbers should represent our best estimate of the amount of time to allow for wait, move, and planned queue time at each work center. Then we can show the sequence of the operations as we expect the order to progress through the routing from the order start date until its due date, as shown in Figure 9–3.

We have therefore created our planned lead time, showing us the order is due on 7/24. The planned lead time is three weeks, so the planned order release should occur on 7/1. In Figure 9–3, we show each of the work operations spread out through that lead time. Technically, we can calculate the planned lead times by taking a normal order quantity and adding together all the operation times plus all the planned queues

FIGURE 9–3
Capacity Requirements Planning (backward scheduling of planned lead time)

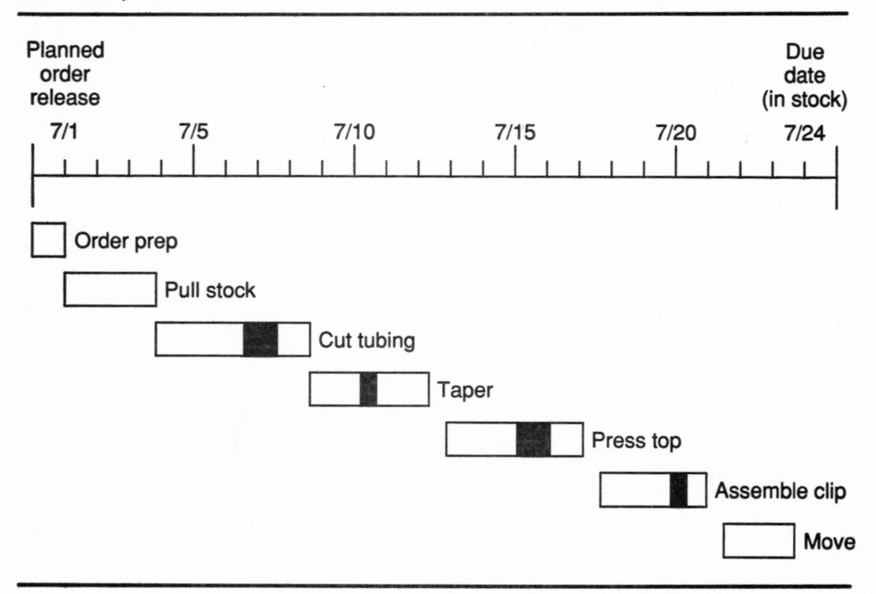

at each operation, plus all the other interoperation times, to arrive at our MRP planned lead time.

Obviously, we can infer that if the order quantity changes, the operation time will change, causing a slight change in the lead time. However, the single greatest element of that lead time will remain the planned queues, as shown in Figure 9–4, where we illustrate all of the lead-time elements for a typical order.

Given this data, we can then calculate the expected start date and finish date for each operation in the routing. We now have a detailed listing of which operations must be done at what point in time and how much time we should allow for those operations. We can make these calculations for all the planned orders plus all the scheduled receipts. This information can then be displayed in several ways, the two most common of which are the *dispatch list* and the *work center load profile.*

DISPATCH LIST

The dispatch list is our tool for priority planning. It should be the listing of all the manufacturing orders in their priority sequence. We most often generate the dispatch list for each work center showing the start date and finish date of each operation, as shown in Figure 9–5. We should also include the order due date. This dispatch list is our way of communicating the MRP order priorities to the people on the shop floor.

The dispatch list should also contain detailed information showing which orders are actually released and what materials are available at that work center. In order for the manufacturing people to view the complete load, it should also include the planned orders and firm planned orders that have not yet been released to manufacturing. We will discuss dispatch lists in more detail in Chapter 10.

As the MRP suggests changes in due dates to match the need dates, the planners can agree to changing the due dates on

FIGURE 9–4
Scheduling and Loading

Part number 20–201
Upper barrel

	Rate	Allowed Hours	Schedule	Hours Required: Tool and Die	Setup	Run Time
Order due 7/24/99		30-piece run				
Paperwork		8.0	7/1/99			
Queue time		8.0				
Pull stock		2.0	7/3/99			
Operation 10 cut tubing						
Tool and die preparation		4.0	7/4/99	4.0		
Queue time		8.0				
Setup	1.5	1.5			1.5	
Run time	0.1	3.0	7/6/99			3.0
Wait		1.5				
Move		1.5				
Operation 20 taper end						
Tool and die preparation		4.0	7/9/99	4.0		
Queue time		17.0				
Setup	2.3	2.3			2.3	
Run time	0.14	4.2	7/11/99			4.2
Wait		1.5				
Move		2.0				
Operation 30 press on top						
Tool and die preparation		4.0	7/13/99	4.0		
Queue time		8.0				
Setup	0.5	0.5			0.5	
Run time	0.04	1.2	7/16/99			1.2
Wait		1.5				
Move		1.5				
Operation 40 assemble clip						
Tool and die preparation		0.0	7/18/99	0.0		
Queue time		8.0				
Setup	0.5	0.5			0.5	
Run time	0.12	3.6	7/21/99			3.6
Wait		1.5				
Move		1.0				
Queue		8.0				
Credit to stock		4.0	7/23/99			
Total lead time	Hours	111.8	Hours	12.0	4.8	12.0
			Percent	11%	4%	11%
				Queue time		
			Hours	57.0		
			Percent	51%		

orders. As they communicate those changes to the MRP, the order changes its location on this priority dispatch list. We should print this list frequently so that the shop people can use it to assist them in deciding which order to run next at their work center.

At the same time, we can take this information and use it to create a work center load profile, as shown in Figure 9–6. Here we show, for each period, the hours of work, called the *load*. The load is made up of the released load, which is the scheduled receipts, plus the planned load, which represents the planned orders. It also should include what we have called *unplanned load*, a time allowance for such activities as preventive maintenance, emergency customer orders, or rework. It is important to keep in mind that the load represents the amount of work that must go through this work center in order to build all the parts supporting our master production schedule. The location, by time period, of this load is determined by the due dates of the orders combined with our estimate of the queue time at the work centers. The purpose of this load profile is to communicate to the people on the shop floor how much work is to be done.

In addition to the load profile, we can print out the listing of the orders due, sorted by work center, and time period. This is called a capacity requirements planning report, as shown in Figure 9–7. Note that in both Figure 9–6 and Figure 9–7, there is a small amount of work indicated in the Past Due column. Ideally, that quantity should be zero, but in most manufacturing companies, there is a planned queue. Consequently, the capacity required indicates how much work must pass through the work center in a given period, but it is not planned that we clean up the total load in any one period.

Now the question becomes, "Are we capable of producing this amount of work in each time period?" To answer this question, we turn to the subject of capacity requirements planning.

FIGURE 9–5
Priority Control Dispatch List

Assemble erasers
Work center: D-02

				Dates					
Work Order	*Part Number*	*Quantity*	*Operation*	*Operation Start*	*Finish*	*Order Due*	*Setup*	*Run*	*Current Location*
39042	18-123	600	020	5/12	5/13	5/14	2.0	7.5	***
37577	18-345	120	020	5/12	5/13	5/14	1.0	8.0	***
30819	18-245	25	020	5/15	5/16	5/17	2.0	8.0	E-91
37576	18-234	130	030	5/16	5/17	5/18	5.0	2.5	***
	Total	875					10.0	26.0	
39272	18-456	450	020	5/17	5/18	5/19	0.8	7.2	***
31569	18-689	1,000	020	5/18	5/19	5/20	2.5	7.5	E-38
38543	18-373	500	020	5/19	5/20	5/21	1.0	4.0	E-33

41383X	18-999	2	040	5/21	5/22	5/23	2.0	0.1	***
38810	18-246	75	020	5/21	5/22	5/23	0.2	1.7	E-12
	Total	2,027					6.5	20.5	
38543	18-373	500	020	5/24	5/25	5/26	1.0	8.0	E-30
39043	18-123	800	020	5/26	5/27	5/28	2.0	10.0	Planned
39146	18-179	300	030	5/28	5/29	5/30	0.5	1.0	***
39012	18-135	1,000	020	5/29	5/30	5/31	0.5	5.0	***
	Total	2,600					4.0	24.0	
38810	18-246	750	020	6/1	6/2	6/3	1.0	9.0	***
39013	18-135	1,000	020	6/1	6/2	6/3	0.5	5.0	E-33
31540	18-783	500	030	6/1	6/2	6/3	1.0	7.5	Issue
31648	18-183	200	020	6/3	6/4	6/5	0.5	2.0	Planned
31649	18-183	200	020	6/3	6/4	6/5	0.5	2.0	FPO
	Total	2,650					3.5	25.5	

*** = Available at work center.

FIGURE 9–6
Load Profile (released and planned load)

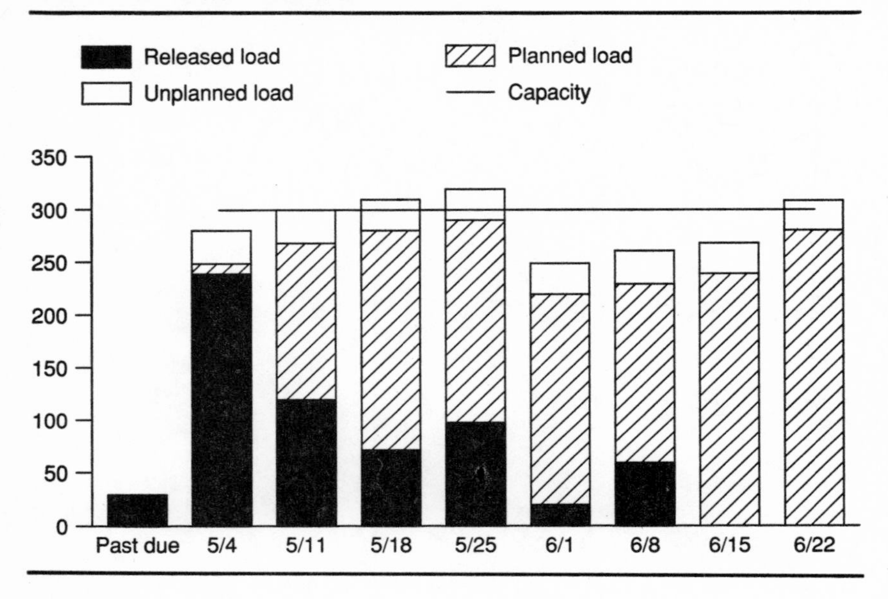

CAPACITY PLANNING

Capacity refers to output potential. Available capacity is the capability to produce output that might best be represented as our demonstrated output in the recent past. Required capacity, on the other hand, represents the capability needed to make a given product mix. Capacity should not be confused with load, which you will remember is the amount of work to be done. The classic way of viewing capacity is to picture the flow of work through the work center as though it were the flow of liquid through a funnel, as shown in Figure 9–8. In this funnel, there is a certain rate at which work is input into the wide end, and a certain rate at which the work is output through the smaller end. Capacity planning is the process of determining what those rates should be. Of course, from a manufacturing point of view, it is desirable to have the rate of output remain the same in

FIGURE 9–7
Capacity Requirements Plan

Assemble erasers
Work center: D-02

	Past Due	Week of 5/12	5/19	5/26	6/3
Capacity:					
Required	10	36	27	28	29
Available		40	40	32	40
Percent		90%	68%	88%	73%

Part Number	*Description*	*Work Order*	*Quantity*	*Due Date*	*Setup Time*	*Run Time*
18-123	Eraser assembly	39042	600	5/14	2.0	7.5
18-345	Eraser assembly	37577	120	5/14	1.0	8.0
18-245	Eraser assembly	30819	25	5/17	2.0	8.0
18-234	Eraser assembly	37576	130	5/18	5.0	2.5
		Total	875		10.0	26.0
18-456	Eraser assembly	39272	450	5/19	0.8	7.2
18-689	Eraser assembly	31569	1,000	5/20	2.5	7.5
18-373	Eraser assembly	38543	500	5/21	1.0	4.0
18-999	Eraser assembly	41383X	2	5/23	2.0	0.1
18-246	Eraser assembly	38810	75	5/23	0.2	1.7
		Total	2,027		6.5	20.5
18-373	Eraser assembly	38543	500	5/26	1.0	8.0
18-123	Eraser assembly	39043	800	5/28	2.0	10.0
18-179	Eraser assembly	39146	300	5/30	0.5	1.0
18-135	Eraser assembly	39012	1,000	5/31	0.5	5.0
		Total	2,600		4.0	24.0
18-246	Eraser assembly	38810	750	6/3	1.0	9.0
18-135	Eraser assembly	39013	1,000	6/3	0.5	5.0
18-783	Eraser assembly	31540	500	6/3	1.0	7.5
18-183	Eraser assembly	31648	200	6/5	0.5	2.0
18-183	Eraser assembly	31649	200	6/5	0.5	2.0
		Total	2,650		3.5	25.5

each time period. The rate of output is determined by the capabilities of our people to produce. Therefore, the capacity, which is analogous to the neck of the funnel, is a measure of our capable rate of output.

FIGURE 9–8
Capacity

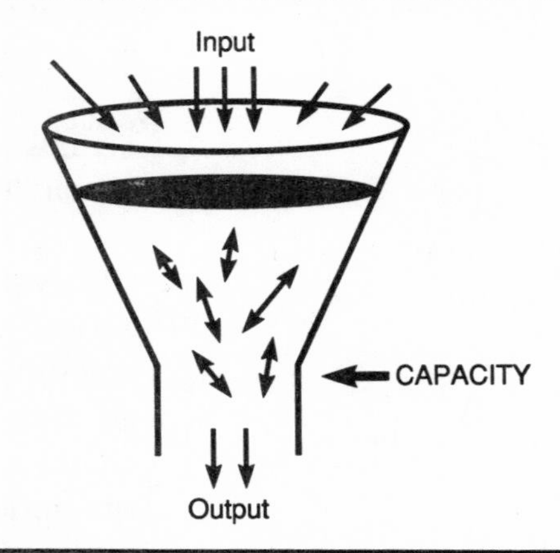

You will remember that when we discussed the flow of work coming through the work center, we mentioned that it is not usual for work to begin on that job immediately—that normally there is a queue in which the job must wait. In our funnel, that queue is shown as the load indicated in Figure 9–9. In some manufacturing shops, this is also known as work in process (WIP), past due, or backlog.

If the input exceeds our output, the load will grow. As the load, or queue, grows, shop lead times tend to grow also. We have already indicated in several places that lead time should be as short as possible, so the queue should be as small as possible. We need to raise the output to a greater rate than the rate of input until we can reduce the load, or queue, to as low a level as possible. This also has the effect of shortening the planned lead times, as described in Focus 9–A.

Capacity requirements planning is planning the required rates of input and output. Here, we should make a distinction between capacity *planning* and capacity *control*. Planning is establishing what those rates of flow are, whereas capacity control is taking corrective action when the rate of output does not match

FIGURE 9–9
Load

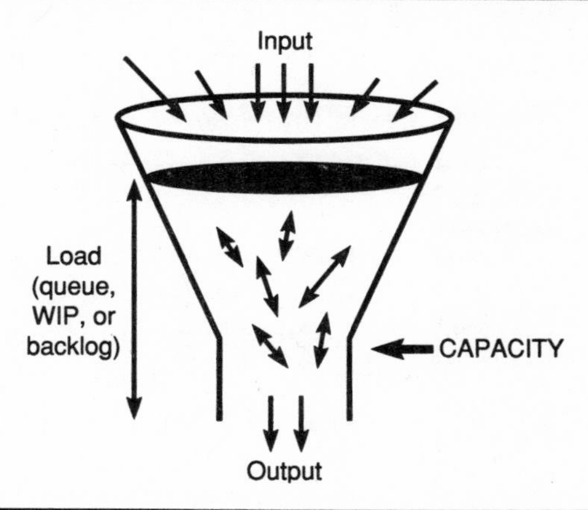

the rate of input. We really want to leave the rate of input alone, because it is determined by the master production schedule, which in turn is determined by the rate at which customers order our products. What we need to focus on is the rate of output. We need to develop the ability to flex our capacity to the extent that we will be able to match it as nearly as possible to the rates of input. (See Focus 9–B.)

MEASURING CAPACITY RATES

Let us examine the rate of output. There are three main ways to calculate that rate. First is *demonstrated capacity,* which is the actual output performance, as verified by data. Most often, it is represented as average standard hours per week.

Next is *rated capacity,* which is a calculated rate of output, in which we take into account the number of machines and the number of people operating those machines, the number of shifts that this work center is operating, the number of hours per shift, and the number of days per week. In Figure 9–10, we

FOCUS 9–A
Focus for Success

To keep *lead times* short—keep *queues* small.

show how many hours the machines are manned, giving us a standard rate of work equal to 160 standard hours per week. In the real world, however, manufacturing operations are not productive 100 percent of the time. Productivity is decreased by two factors, utilization and efficiency. Utilization is the percentage of the time a work center is actually producing output, as compared to the total clock time that the work center is staffed. There are all sorts of activities that occur in a manufacturing environment that will cause a work center to stop producing: machine downtime, clean-up time, employee meetings, and other activities.

In Figure 9–10, utilization of 90 percent means that for 160 clock hours per week, the machine is working, on either setup or run, for only 144 hours.

Efficiency, on the other hand, is a measure of how well our operators are able to perform compared to the standard hours set on our routing. Some very experienced operators may be able to produce the parts in less time than the standard dictates; they are considered to be very efficient. An inexperienced oper-

FOCUS 9–B
Focus for Success

Focus on the *rate of output* and flex it to match the *rate of input*.

FIGURE 9–10
Rated Capacity

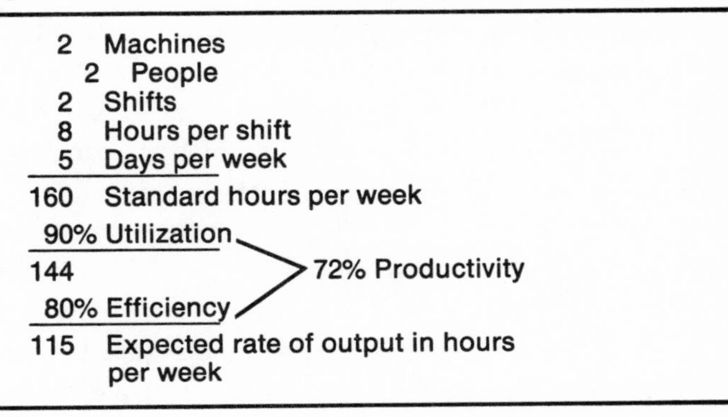

ator, on the other hand, or one who has problems with the job, and who therefore takes more than the standard time, would have an efficiency of less than 100 percent. Efficiency is defined as the standard hours earned divided by the actual hours worked. The standard hours earned is the good pieces produced multiplied by the standard hours per piece as recorded in the routing.

2 Machines
2 People
2 Shifts
8 Hours per shift
5 Days per week
160 Standard hours per week
90% Utilization
144
72% Productivity
80% Efficiency
115 Expected rate of output in hours per week

In Figure 9–10, we show 80 percent efficiency; therefore, we can expect 115 hours worth of standard work to be produced each week at this work center. This is our real expected rate of output given our current manning conditions.

The third capacity measure is *theoretical maximum capacity,* which is the capacity allowing no adjustments for planned downtime or shutdowns, while we are running three full shifts with no allowance for lunches or breaks. In our example, this would give us a theoretical maximum capacity of 240 hours per week.

Capacity planning should establish our planned output rate. Most companies look at the rated capacity. On the other hand, the rate at which work is input to the work center is defined by the MRP requirements to support the master schedule. Our MRP continues to produce planned orders and scheduled receipts that create the input to our work centers. Our capacity requirements planning report therefore shows the work required each week to support the master production schedule. Capacity control is comparing those rates of input to the rates of output and taking some corrective action. This is the most troublesome spot in an MRP system, because if we continue to plan orders at the master production schedule that create an input that exceeds the output rate, our backlog will continue to grow. The assumption in an MRP system is that the capacity requirements planning report will show the required amount of output, no matter what the rated capacity is. This characteristic is called *infinite loading,* and we will examine it in Chapter 11.

The premise is that management will exercise control over capacities by adjusting the rate of output. We will explore some of the techniques for doing this in Chapter 11 also.

APPLICATIONS

Many companies today feel that capacity is limited to the rated capacity calculation, but when we look at world-class companies, we see them moving toward the process industry concept, in which productive equipment should run nonstop, 24 hours a day, seven days a week. It should only be stopped for required preventive maintenance. We have seen one high-volume batch-processing company increase its production by 20 percent by working through lunches and breaks. The company did this by staggering operator breaks, and the real gain came not so much from the time as it did from the improvement in process dependability when the shutdown and startup of the line three times each shift was eliminated.

But today's world-class companies don't carry this to extremes, producing constantly just for the sake of producing. It is more important to be able to flex the rate of output to match the rate of input. If our customer orders create a slow input rate, we should be able to slow down the output. On the other hand, if the customer demands increase the input, we should have the flexibility to call in more people or whatever else it takes to increase the output rate. Permanent part-time workers are a viable alternative in achieving this goal.

One company we know of, for example, will call in part-time workers late in the day, as many as are needed for that specific day's work, to complete that night's scheduled output.

Another manufacturing firm has college students lined up to come in and run machines on an as-needed basis in order to flex its capacity to match the requirements for furniture components.

We should be cautious about creating overcomplex routings with finicky detail concerning work centers, operations, and times. As George Plossl has frequently observed, this creates "an illusion of precision." We don't have to track every individual work center with every individual standard and every individual part. Rather, we should look at the process of manufacturing the part and use the computer system to plan and track aggregate groupings of work centers, such as lines or cells. We know of a company that uses its master scheduling system to plan the overall rate of input and the purchase of raw materials. The company releases small quantity orders to a production line for that product. The people plan the rate of input and output for the line in total and do not try to track the individual rates for each operation in the line. When the company's work centers had individual queues and were spread throughout the shop, the planned lead time to produce a particular family of parts was 90 days. After moving the equipment into a flow line and making other process improvements, the company was able to reduce the queues so that the lead time now is less than 10 days. In this company, the MRP is used only to signal the delivery of raw materials. The planners release replenishment orders for end units, tracking only one work center that represents the full production line. Inside the production line, the company uses visual signals and a pull system to progress the part through the operations.

SECTION 4

MANAGING WITH MRP

CHAPTER 10

MANAGING MRP PRIORITIES

The heart of an MRP system is where the priorities are managed. This is where the material requirements *planning* system is transformed into an execution system. When we refer to *priorities,* we are talking about the sequence in which the parts will flow into our stockroom. The priority establishes the relative importance of the job; it also establishes the sequence in which the jobs should be worked on. Of course, for parts that are manufactured in our own shop, we can establish that sequence in detail for each operation in the routing, whereas for a purchased part, we merely establish a receipt date from the vendor.

Our discussion of priorities will be broken out into three major areas: purchased parts, manufactured parts or components, and finished goods. Before we do this, however, we should consider the two important aspects of priority: *priority planning* and *priority control.*

Planning is the process of determining what is needed and when we need it. This is the same as the objective of the material requirements planning system that is driven by the master schedule. As we have already discussed, the objective of our MRP system is to maintain the proper due dates on required materials. That is a basic definition of priority planning, as well.

Priority control, on the other hand, is the process that we use to actually execute our plan. The primary vehicle for doing this is the dispatch list. You will remember that the dispatch list is a listing of the manufacturing orders in priority sequence. Control implies that we use this dispatch list to make actual decisions and direct people on the shop floor to work on specific orders in specific sequence. We sometimes refer to this process as dispatching, which means assigning these orders to specific workers. In today's modern manufacturing organization, this dispatching function, which used to be the purview of the dispatcher, is now often performed by the shop supervisor or team leader and is sometimes communicated directly to the operators via computer screens or reports.

DISPATCH LIST

A typical dispatch list from FineLine Industries is shown in Figure 10–1. (This is a duplicate of Figure 9–5.) This is a listing of all the work that will cross our work center to assemble erasers. What we are actually doing here is attaching the little metal eraser holder, or socket, to the rubber eraser. As we have defined the dispatch list, it is a listing showing work order number, part number, and quantity due. The real instrument of the priority control is the start date, defining when the work should begin on this work order. As you can see, the dispatch list is sorted in order by start dates, allowing us to readily see the sequence in which the work should be performed. We also see the date on which the operation should be finished, and for additional information, the date on which the order is due into our stockroom. The dispatch list also incorporates capacity information, such as the setup hours and run hours, to help us evaluate the time it will take to complete each work order. Most dispatch lists also include helpful supplemental information, such as remarks and current location. In our example, the current location shows where the material is located for each work order. It is important to know whether the materials have actually been delivered to the work center and are available for the operators to work on. In Figure 10–1, that status is shown by three asterisks (***). This category represents available work that is actually in the department and can be performed, as opposed to work that is scheduled but not yet on hand. Scheduled work can be of two types—work in the shop but still resident at a previous operation (in our example, the location column shows the number of the work center where the work is now). The other type is work that is in the planned state, which means that it is a planned order release on our MRP worksheet. This is future work that has not yet been released to our shop floor. All of this is consistent with our MRP worksheet for erasers (Figure 8–2), which is shown here as Figure 10–2. The scheduled receipt in Figure 10–2 for the 600 pieces shows up on our dispatch list as Work Order 39042, showing that the 600 pieces should be started on 5/12, will take 7.5 hours to assemble, and should be put into stock by 5/14. Notice also that there is a planned order release for 800 pieces that should be released for completion to stock by 5/28. Therefore, it shows up on our dispatch list to be started on 5/26. That line shows it to be in planned status. We can see that the dispatch list is an accurate reflection of the priorities established in the MRP planning process. It also reflects the total amount of work that is both planned and released to our shop floor. Some companies incorrectly list only the available work on their dispatch lists, assuming that the operators are not interested in the work coming up but only in the work available right now. As we have mentioned before, one of our objectives is the reduction of queues on the shop floor, which has the psychological effect on the operators of leading them to believe that they are going to run out of work. The dispatch list, in showing planned work, gives the operators a view of the work that is coming down the pipeline.

As order priorities change, the dispatch list must change also. In Figures 8–3 and 8–4, we changed the master schedule, causing the MRP to generate a reschedule notice. In Figure 8–4, which is reproduced as Figure 10–3, the material planner for the eraser implemented the suggestions from the MRP system. You recall that the planner rescheduled Order 39042 for 600 pieces out to Period 6/3, and she also released the new Order 39043 for 800 pieces, also due 6/3. The result of these actions would appear on a revised dispatch list as shown in Figure 10–4. This is how we tie together the dynamics of priority planning with the

FIGURE 10–1
Priority Control Dispatch List

Assemble erasers
Work center: D-02

				Dates					
Work Order	*Part Number*	*Quantity*	*Operation*	*Operation Start*	*Finish*	*Order Due*	*Setup*	*Run*	*Current Location*
39042	18-123	600	020	5/12	5/13	5/14	2.0	7.5	***
37577	18-345	120	020	5/12	5/13	5/14	1.0	8.0	***
30819	18-245	25	020	5/15	5/16	5/17	2.0	8.0	E-91
37576	18-234	130	030	5/16	5/17	5/18	5.0	2.5	***
	Total	875					10.0	26.0	
39272	18-456	450	020	5/17	5/18	5/19	0.8	7.2	***
31569	18-689	1,000	020	5/18	5/19	5/20	2.5	7.5	E-38
38543	18-373	500	020	5/19	5/20	5/21	1.0	4.0	E-33

41383X	18-999	2	040	5/21	5/22	5/23	2.0	0.1	***
38810	18-246	75	020	5/21	5/22	5/23	0.2	1.7	E-12
	Total	2,027					6.5	20.5	
38543	18-373	500	020	5/24	5/25	5/26	1.0	8.0	E-30
39043	18-123	800	020	5/26	5/27	5/28	2.0	10.0	Planned
39146	18-179	300	030	5/28	5/29	5/30	0.5	1.0	***
39012	18-135	1,000	020	5/29	5/30	5/31	0.5	5.0	***
	Total	2,600					4.0	24.0	
38810	18-246	750	020	6/1	6/2	6/3	1.0	9.0	***
39013	18-135	1,000	020	6/1	6/2	6/3	0.5	5.0	E-33
31540	18-783	500	030	6/1	6/2	6/3	1.0	7.5	Issued
31648	18-183	200	020	6/3	6/4	6/5	0.5	2.0	Planned
31649	18-183	200	020	6/3	6/4	6/5	0.5	2.0	FPO
	Total	2,650					3.5	25.5	

*** = Available at work center.

Figure 10–2
MRP Worksheet (eraser assembly)

Lead time = 5

	Now	1 4/23	2 4/30	3 5/7	4 5/14	5 5/21	6 5/28	7 6/3	8 6/10
Requirement			1,600		1,000		800		
Scheduled receipts					600				
On hand	2,000	2,000	400	400	0	0	0	0	0
Planned order releases		800							

Period

Suggestion: Scheduled receipt: Order 39042 for 600 due 5/14
New order for 800 due 5/28 Release now

Figure 10–3
MRP Worksheet (eraser assembly)

Next Week
Lead time = 5

Period

	Now	1 5/7	2 5/14	3 5/21	4 5/28	5 6/3	6 6/10	7 6/17	8 6/24
Requirement			1,000		800	1,600			
Scheduled receipts						600 800			
On hand	2,000	2,000	1,000	1,000	200	0	0	0	0
Planned order releases		800							

Schedule receipt: Order 39042 for 600 due 6/3
Schedule receipt: Order 39043 for 800 due 6/3

MRP worksheet to control priority on the dispatch list. The dispatch list communicates the order priorities to the shop floor.

SHOP FLOOR CONTROL

Let's think about the shop floor for a minute. We have to remember that the folks there are interested in the other conditions that are required to produce the order: tooling, the combination of setups for like items, and so on. These are typically called *operation priorities*. The shop is always interested in maximizing production through the examination of the operating priorities, and frequently shop people will look over the dispatch list to "cherry pick" jobs that can be grouped together. On the other hand, the MRP planning process communicates the *order priorities* that are necessary in order to meet the master production schedule. The dispatcher's role is to weigh these alternatives and decide which is in the best interests of the entire company. Occasionally we have situations in which the order priority and the operation priority coincide, and the jobs can be produced with the same setup, as in our eraser example.

More often, a compromise must be made between order and operation priorities, so we would like to see more detail included in our dispatch list. Figure 10–5 is an example of such a dispatch list, showing us as additional operation information the tool numbers. The dates shown on the dispatch list should tie into the lead times established on the routing. This is the foundation of the MRP planned lead times. As shown in Figure 10–6, the start date represents the date on which the operation should start, and the end date shows the date on which the operation should be completed. The due date is the date on which the order is due into stock. The planned order release date is the date on which the order is released to the shop and is the date on which the components should be available to pull from stock. Here is the point at which the planning system and the execution are tied together.

FIGURE 10–4
Priority Control Dispatch List (assembly)

Assemble erasers
Work center: D-02

				Dates					
Work Order	*Part Number*	*Quantity*	*Operation*	*Operation Start*	*Finish*	*Order Due*	*Setup*	*Run*	*Current Location*
37577	18-345	120	020	5/12	5/13	5/14	1.0	8.0	***
30819	18-245	25	020	5/15	5/16	5/17	2.0	8.0	E-91
37576	18-234	130	030	5/16	5/17	5/18	5.0	2.5	***
	Total	275					8.0	18.5	
39272	18-456	450	020	5/17	5/18	5/19	0.8	7.2	***
31569	18-689	1,000	020	5/18	5/19	5/20	2.5	7.5	E-38
38543	18-373	500	020	5/19	5/20	5/21	1.0	4.0	E-33
41383X	18-999	2	040	5/21	5/22	5/23	2.0	0.1	***

38810	18-246	75	020	5/21	5/22	5/23	0.2	1.7	E-12
	Total	2,027					6.5	20.5	
38543	18-373	500	020	5/24	5/25	5/26	1.0	8.0	E-30
39146	18-179	300	030	5/28	5/29	5/30	0.5	1.0	***
39012	18-135	1,000	020	5/29	5/30	5/31	0.5	5.0	***
	Total	1,800					2.0	14.0	
39042	18-123	600	020	6/1	6/2	6/3	2.0	7.5	***
39043	18-123	800	020	6/1	6/2	6/3	2.0	10.0	Issued
38810	18-246	750	020	6/1	6/2	6/3	1.0	9.0	***
39013	18-135	1,000	020	6/1	6/2	6/3	0.5	5.0	E-33
31540	18-783	500	030	6/1	6/2	6/3	1.0	7.5	Issued
31648	18-183	200	020	6/3	6/4	6/5	0.5	2.0	Planned
31649	18-183	200	020	6/3	6/4	6/5	0.5	2.0	FPO
	Total	4,050					7.5	43.0	

*** = Available at work center.

FIGURE 10–5

Dispatch List

Work center: A30
Cut tubing

Work Order	Item Number	Operation Number	Description	Tool Number	Dates			Quantity Due	Hours	
					Start	End	Due		Setup	Run
31328	20-201	010	cut tubing	A-63782	7/4/99	7/8/99	7/24/99	30	1.5	3.0
31339	20-217	010	cut body		7/6/99	7/9/99	7/21/99	25	1.0	4.0
31342	20-391	010	cut tube	A-61751	7/9/99	7/11/99	7/19/99	50	1.7	6.2

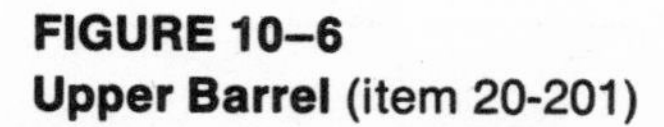

FIGURE 10–6
Upper Barrel (item 20-201)

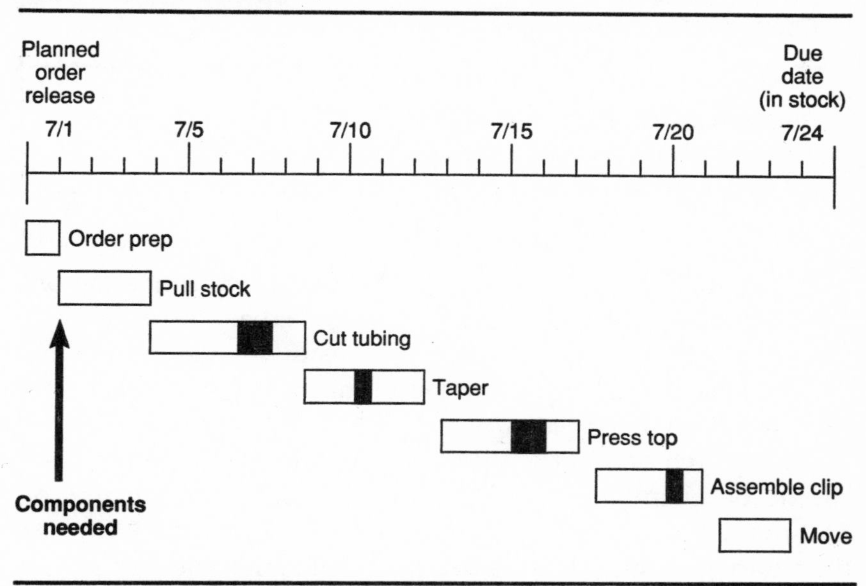

DUE DATE MUST MATCH NEED DATE

MRP gross-to-net logic creates detailed plans that will suggest that the shop order due dates need to match the requirements need dates. This is the foundation principle of an MRP system, as shown in Focus 10–A. The MRP will always suggest that these dates stay matched, alerting the planner to investigate the feasibility of implementing those suggestions. The first objective, remember, is to "move heaven and earth" to be able to meet those need dates. As our team agrees that we can meet the schedule, we will advise the system of actual shop order due dates, which will show up on the MRP worksheet and on the dispatch list. The shop supervisor uses the dispatch list as the integration of order priorities with operation priorities. The dispatch list thereby becomes the vehicle by which our plan is executed.

FOCUS 10–A
Focus for Success

Shop order *due* dates
Must match requirement *need* dates.

Back in Chapter 7, we stated that manufacturing shop orders should not show the due date because that date is likely to change as order priorities change. Instead, we should use the dispatch list as the vehicle for communicating to our people on the shop floor the most current date for shop orders.

For purchased parts, the same priority planning emphasis should be maintained. This is typically done with a listing of part numbers and their due dates for each vendor, as shown in Figure 10–7. Notice the due date is the date that the part is due in our stockroom, available to fill requirements. Our MRP planning system will suggest that these due dates match up with our need dates in the same way they will for manufactured parts. We do not need a detailed routing, obviously, for purchased parts, but we should take particular note that there are certain elements of lead time for a purchased part that should be incorporated into the MRP planned lead time. Figure 10–8 shows that we should make allowance for the time it takes to do the paperwork and release the purchase order to the vendor; the vendor's own lead time; the transit time from the vendor's facility to our own; and our own receiving, inspection, and stocking time. It is critical to consider all of these elements in the establishment of the MRP planned lead time because the purchase order planned due date must match our need date and that date has to be the date when the material is in stock and actually available for our use.

The equivalent of the dispatch list, the purchase order due date list, shown in Figure 10–7, should be used by the vendor to

FIGURE 10–7
Purchase Due Dates

Supplier: Jones Steel

Vendor Part No.	*Quantity*	*Description*	*Due Date*	*Vendor Ship*
3791	300 feet	Tubing	6/28/99	6/27/99
3851	250 feet	Tubing ⅛" diameter	6/30/99	6/29/99
417	12 feet	Channel 3 × ¼"	7/1/99	6/30/99

sequence the work flowing from its facility to ours. As our requirements change, the planners will see new need dates and communicate these requested changes to the vendor. When these changes are agreed on, the new need dates will be entered into the system, causing the purchase orders to be sequenced into different priorities.

FINISHED GOODS PRIORITIES

Now we can turn our attention to managing the priorities of our finished goods. We have several alternatives available to us. Our first choice could be to plan a master schedule for each end unit, considering each part to be a separate item, discrete from

FIGURE 10–8
Purchase Lead Time

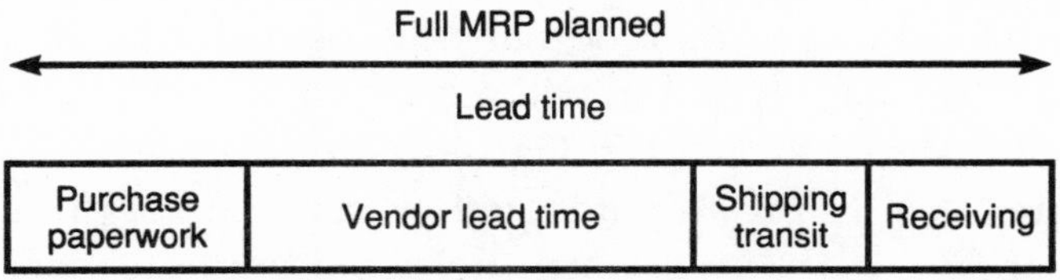

Full MRP planned
Lead time

Purchase paperwork	Vendor lead time	Shipping transit	Receiving

all others, in which we establish inventory management policies similar to the sawtooth pattern described in Chapter 2. Each item would be managed with its own safety stock, establishing priorities based on the backlog. We would use the technique for master production scheduling described in Chapter 6. As customer requirements change and our own organization changes its forecasts, the planned releases for the master schedule items would be changed, driving the component priorities in their turn. This is a relatively straightforward process, but it requires that a person look at every item on a regular basis. This can be very time-consuming, particularly if we consider the proliferation of product options and variations required by today's modern manufacturing organization.

An alternative method would be to utilize the planning bill concept as described in Chapter 4. Here, we are forecasting at the planning bill—or family—level, but the priorities are being managed for each individual component number.

The best way to manage the end-unit priorities is to not stock the end units at all but rather to stock components. Remember that a component can be a raw material, ingredient, part, or subassembly that goes into a higher-level assembly or end item. In order to effectively manage the supply and demand, as well as the priority of these parts, we utilize two techniques, *two-level master scheduling* and *final assembly scheduling*.

TWO-LEVEL MASTER SCHEDULING

First, we will examine two-level master scheduling. Here, we utilize the technique of modular bills of material, which we discussed in Chapter 4. You will recall that we created a modular bill of material for pens in Figure 4–16, which is reproduced as Figure 10–9. In order to adequately manage the priorities of the components that make up the pens, we must merge the modular bill of material with the master schedules. A planning

FIGURE 10–9
Modular Bill ("Pens")

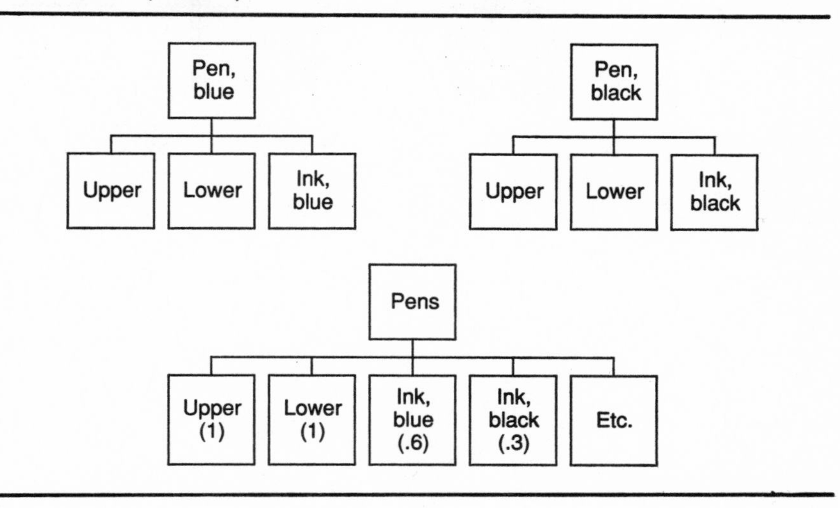

bill or modular bill is used to master schedule a family along with key options and features. In our example, the planning bill is for pens, where each pen is made up of one upper assembly and one lower assembly. We had estimated that 60 percent needed blue ink cartridges and 30 percent needed black ink cartridges. There are other options and features, of course, but we'll keep our example simple and consider only these two choices. Let us start with a situation where we have no customer orders and see how we can use the forecast to plan the dependent priorities. As shown in Figure 10–10, we take the forecasted pen sales and multiply each forecast by the corresponding percentage to arrive at the forecast for the ink cartridges. As you can see in Period 1, our forecast for 2,000 pens generates a forecast for 600 black ink cartridges and 1,200 blue ink cartridges. The rest can be for other colors that would have their corresponding master schedules. The upper barrels and the lower barrels each have their own master schedule sheets. Using these forecasts, we can now plan the master production schedule quantity for each ink cartridge and use them to drive

FIGURE 10–10
Two-Level Master Scheduling

Level Zero

Pens, planning bill 42-377

	1	2	3	4	5
Forecast	2,000	2,000	2,000	2,000	2,000
orders	0	0	0	0	0
On hand	0	0	0	0	0
Available	2,000	2,000	2,000	2,000	2,000
* MPS *	2,000	2,000	2,000	2,000	2,000

Explode available-to-promise to next level forecast.

Level One

First
30.0% Option black ink cartridge Part 37-7215

	1	2	3	4	5
Forecast	600	600	600	600	600
orders	0	0	0	0	0
On hand	50	100	150	200	250
Available	650	650	650	650	650
* MPS *	650	650	650	650	650

Second
60.0% Option blue ink cartridge Part 37-7213

	1	2	3	4	5
Forecast	1,200	1,200	1,200	1,200	1,200
orders	0	0	0	0	0
On hand	100	200	300	400	500
Available	1,300	1,300	1,300	1,300	1,300
* MPS *	1,300	1,300	1,300	1,300	1,300

our MRP system. As changes are made in the forecast, this double level of master scheduling would indicate that we should make changes in the master schedule quantities, thereby changing the priorities of the lower-level components. We all know, however, that the most common cause of change in those priorities is changes in the mix of actual customer orders, rather than changes to the master schedule.

We turn our attention to entering the customer orders using modular bills of material. This is commonly known as *configuring a customer order*. Here, when the customer orders an assembly, we are able to pick and choose the options and features the customer wants. Most often, from that selection process we create a temporary bill of material for this unique combination of parts. This type of bill is frequently stored as a temporary bill that exists for the duration of the order and is then dropped from the database. The most prevalent example of this is in the automotive industry, but it is present in many environments where the customer is offered a wide variety of options and features. A configuration menu for a custom order for a pen might look similar to the diagram in Figure 10–11, in which case we can select a pen with an upper barrel, a lower barrel, and whatever color ink cartridge the customer wants. Three things will then happen as a result of this process of configuring a customer order. First, a unique bill of material is created, defining the exact configuration that the customer wants at this time. Second, a work order for this assembly is loaded into the *final assembly schedule*. The final assembly schedule is where the operations are planned to complete the product from the level at which it is stocked (the master schedule level) to the end-item level. In our example, we take an upper barrel, a lower barrel, the appropriate ink cartridge, and whatever packaging is required, to fill the customer order.

FIGURE 10–11
Configuration (Pen)

- Upper barrel
- Lower barrel
- Ink cartridge
    - Blue
    - Black
    - Red
    - Other

Rather than try to discretely forecast every possible combination of options that the customer might want, we plan to run the MRP system by building only the components and then assembling to customer orders. Therefore, our master production schedules in the MRP system must reflect the requirements and the priorities of this changing need for specific options. The third thing that is done for a configured order, then, is that we consume the forecast of the planning bill and allocate actual customer orders to the component master schedules. This can be illustrated by an example such as the one in Figure 10–12. We enter a customer order for 100 pens with black ink cartridges, showing that orders for 100 have been entered for the pen, with 100 orders for black ink cartridges and none for the blue ink cartridges. Before we look at all the other customer orders that are involved, let us go through the process of two-level master scheduling to see how it works.

First, we establish a master schedule for the modular bill for the pen. The forecast is for 2,000 pens for Period 1, and we ask ourselves if there is any special way we should group the manufacture of the components. We never really *build* the thing called the "pen," but we are establishing the priorities for the matched sets of components. Many times, what will happen in the case of the 2,000 forecast is that we will say that the master schedule is 2,000. Of course, one customer order for 100 leaves the available-to-promise figure at 1,900. Now we can explode one level in the bill of material, the available-to-promise number that is used as the basis for the requirements for the components. The forecast for black ink cartridges would show 570 required (30 percent of 1,900), and 1,140 blue ink cartridges needed (60 percent of 1,900). Looking at each master schedule, we see that the total needed for black ink cartridges is now 670, so we can set our production quantity accordingly. We can do the same thing for each of the other components—blue ink cartridges, upper and lower barrels—that make up what we call the pen.

This is also spread across multiple time periods, very nicely showing customer orders and forecasts, as in Figure 10–13.

FIGURE 10–12
Two-Level Master Scheduling with a Customer Order

Level Zero

Pens, planning bill 42-377

	1	2	3	4	5
Forecast	2,000	2,000	2,000	2,000	2,000
orders	100	0	0	0	0
On hand	0	0	0	0	0
Available	1,900	2,000	2,000	2,000	2,000
* MPS *	2,000	2,000	2,000	2,000	2,000

Explode available-to-promise to next level forecast.

Level One

First
30.0% Option black ink cartridge Part 37-7215

	1	2	3	4	5
Forecast	570	600	600	600	600
orders	100	0	0	0	0
On hand	0	50	100	150	200
Available	550	650	650	650	650
* MPS *	670	650	650	650	650

Second
60.0% Option blue ink cartridge Part 37-7213

	1	2	3	4	5
Forecast	1,140	1,200	1,200	1,200	1,200
orders	0	0	0	0	0
On hand	160	260	360	460	560
Available	1,300	1,300	1,300	1,300	1,300
* MPS *	1,300	1,300	1,300	1,300	1,300

There are a couple of things that we should keep in mind about this process of two-level master scheduling. One, of course, is that we can have a balance on hand of components, so that we can compute the projected balance on hand by adding the master schedule, subtracting the forecast, and subtracting customer

FIGURE 10–13
Two-Level Master Scheduling with Several Customer Orders

Level Zero

Pens, planning bill 42-377

	1	2	3	4	5
Forecast	2,000	2,000	2,000	2,000	2,000
orders	100	100	50	50	0
On hand	0	0	0	0	0
Available	1,900	1,900	1,950	1,950	2,000
* MPS *	2,000	2,000	2,000	2,000	2,000

Explode available-to-promise to next level forecast.

Level One

First

30.0% Option black ink cartridge Part 37-7215

	1	2	3	4	5
Forecast	570	570	585	585	600
orders	100	0	17	25	0
On hand	0	80	128	168	218
Available	550	650	633	625	650
* MPS *	670	650	650	650	650

Second

60.0% Option blue ink cartridge Part 37-7213

	1	2	3	4	5
Forecast	1,140	1,140	1,170	1,170	1,200
orders	0	100	33	25	0
On hand	160	220	317	422	522
Available	1,300	1,200	1,267	1,275	1,300
* MPS *	1,300	1,300	1,300	1,300	1,300

orders. Using this comparison of supply and demand for the components, we are able to manage the priorities of these parts. One other point is that, at the modular bill of material level (that is, the pens), we must use the available-to-promise figure for each period; therefore, we cannot use the cumulative available-to-promise figure that our marketing department might prefer to have. We can, however, use the cumulative available-to-promise at the component level, which can be used for customer order promising and what-if scenarios.

The purpose of all of this, of course, is to use the MRP system to drive the major components into the stockroom, without assembling those components into a finished end unit until we learn the specific requirements from the customer's order. We can retain the inventory in a more flexible form and respond to a wider variety of customer needs. We can expect more accurate forecasting, as well, because our forecast is for the aggregate family instead of individual end products. We can say that the percentages used in the modular bill to distinguish black ink cartridges from blue ink cartridges are a commitment to the mix of our various options. It is difficult for companies to put into place a review of these percentages. One technique that is very successful is overplanning, which was discussed in Chapter 4. Remember that overplanning only works for the forecasted amount, not for orders that have actually been entered.

APPLICATIONS

In our discussion of purchasing, we referred to MRP planned lead time and the need to incorporate several elements: vendor lead time, purchasing paperwork, transit, and receiving times. In today's modern business, companies are making great strides toward the reduction of several of these elements. Receiving, for example, is checking materials in, inspecting them, and moving the materials into stock. More and more companies are establishing supplier partnerships, wherein checking and inspection can be eliminated because of proven vendor reliability.

Purchasing paperwork includes transmitting a requisition to purchasing, obtaining bids, and maintaining purchasing files, and can be a very time-consuming process. We can use the visibility of the MRP system to reduce the length of this paperwork processing by showing planned order releases to purchasing and to the supplier prior to the actual planned order release date through the method known as the *blanket order*. Here, we establish a total quantity to be delivered over a given period of time, with multiple releases scheduled for different dates. The ultimate goal is to establish a flow pattern for the material that will allow us to make the batch sizes cover a small increment of time. One company producing hydraulic lifts that require steel tubing receives a truckload delivery of two different sizes of tubing at the plant every morning between 7:00 and 7:30. This quantity is sufficient for a day's work. If and when the material priorities change, the company can respond as quickly as the next day to changes in customer requirements. Of course, the speed and volume of these changes are a significant point in the development of that customer-supplier partnership.

Another area to examine is the concept of two-level master scheduling with a configured bill of material. A manufacturer of hydraulic motors utilizes this concept to great strategic advantage. The company does top-level master scheduling on six sheets of paper, representing the six product families that it manufactures, using the two-level bill of material to manage the priorities of all its components. In addition, the final assembly schedule shows the assembly of matched sets of components three levels deep in the bill of material. Obviously, the manufacturer has flattened the bill of material to enable it to create a dispatch list from its final assembly schedule. There are parts that must be machined to match other parts, and the company uses the list of actual customer specifications to decide how to machine those parts. These completed parts then travel to the final assembly line to be combined in final assembly with standard parts, and are tested, packaged, and shipped. This is an excellent example of a company being able to take the MRP logic and work with it to reduce the levels. This allows it to operate more like a make-to-order than a make-to-stock organization. Through creative application of scheduling principles, a company can retain flexibility in response to customer demands.

CHAPTER 11

MANAGING CAPACITIES

Now that we have addressed the management of MRP priorities and have established order priorities based on customer requirements, we can turn our focus to the execution of our plan. This is done through the process of managing capacities.

You will recall that we discussed capacity requirements planning in Chapter 9, describing capacity as the capability to produce output. We used the funnel in Figure 9–9 (repeated here in Figure 11–1) to illustrate the relationship between inputs, outputs, and load. Capacity requirements planning utilizes scheduled receipts, plus planned order releases, plus firm planned orders to compute the planned rate of input to each work center. *Capacity planning,* therefore, is the process of defining the resources we must have in order to make the output equal the input.

Capacity control, on the other hand, is measuring that rate of output, comparing it to the input, and, if a variance exists, taking corrective action. There are two general categories of corrective actions: Change the rate of input and change the rate of output.

Changing the rate of input implies that we change the amount of product to be built, which in the aggregate sense means cutting off customer orders. This is not an acceptable course of action; rather, we should focus on changing the rate of output. Dealing with this question of how to change the rates of output, we can see that there are multiple levels of detail that should be addressed.

We can think in terms of the whole company as being represented by a funnel, with customer orders as the input

FIGURE 11–1
Capacity

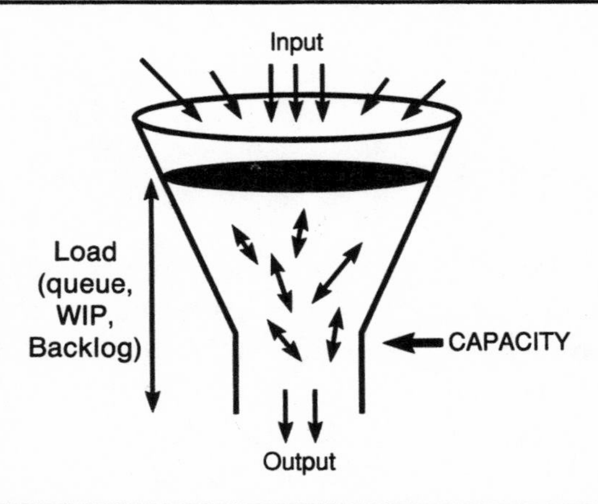

and shipped products as the output. Using this viewpoint, we obviously should not restrict the rate at which work comes into the company; rather, we want to increase the output to match that input. If, on the other hand, we look at the funnel as an individual work center within the company, then it may be practical to talk about changing the rate of input by diverting some of the work to alternate work centers. The real trick in managing capacities is to first take the global view, then refine it as we proceed through a closed-loop plan. This, in fact, is consistent with what we said in Chapter 1 about the planning process as an iterative top-down process that becomes progressively more detailed as we go along.

We will examine five different techniques at four different levels of the planning process, as depicted in Figure 11–2. In the long-range time frame, we will look at *resource requirements planning,* which corresponds to our production planning. We will also look at *rough-cut capacity planning,* which translates our production schedules into capacity requirements. In the medium time range, we will look at *capacity requirements planning,* which translates our detailed MRP into work center requirements. In the very short range, we will look at *input/output*

FIGURE 11–2
The Planning Process

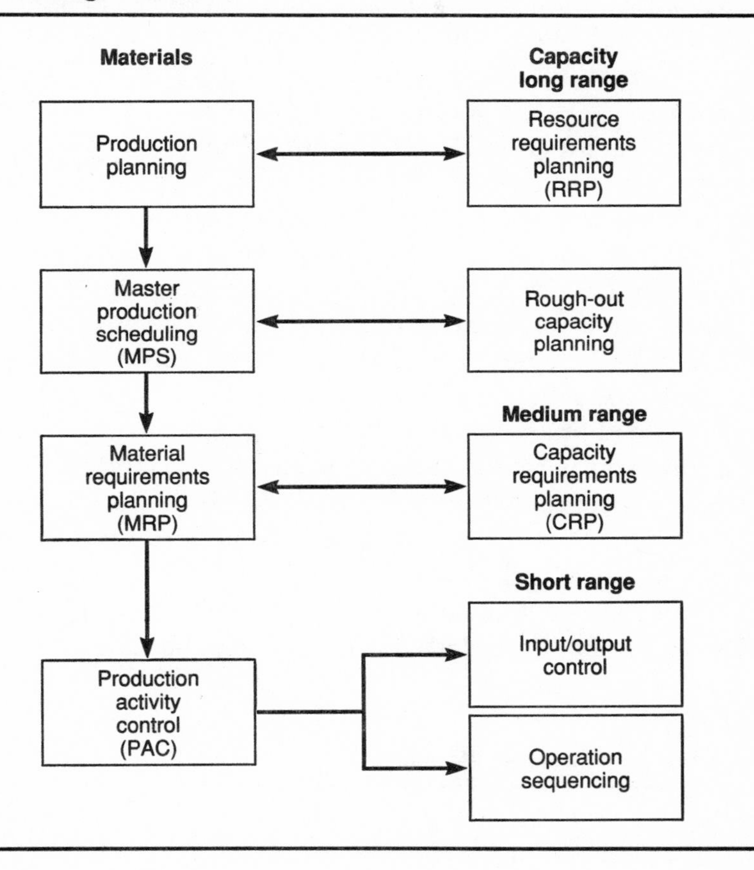

controls, which measure the performance of our production activity control. We will also look at *operations sequencing,* which is a simulation technique used to manage our shop floor execution system.

RESOURCE REQUIREMENTS PLANNING

Resource requirements planning deals with the long-range planning horizon. This refers to items that take a long time to acquire and call for substantial capital outlay, such as land, plants, and machinery. We start with the production plan, which shows the product families and the expected aggregate output of each. We want to derive a method of evaluating our ability to produce the required levels of output. We should measure these levels of output in terms of that critical resource that requires the most time to acquire. In some organizations, that may mean the time that it takes to hire and train new people, so we may drive change with aggregate changes in the staffing level, which can then be projected into plant expansions and equipment acquisitions. A typical time frame for this would be anywhere from 1 year up to 5 or 10 years. The method we use to do this is the aggregate planning concept, in which we take the total production required times a measure of the historical output rate (frequently called the *bill of resources,* which is usually an indirect measure, not a scientifically derived industrial engineering concept). A typical example might be to use average direct labor hours per sales dollar, or standard hours per unit produced, such as the example in Figure 11–3.

Here, we see that in the past six months, we have produced 1,150 of the family called "Pens." We can see the number of

FIGURE 11–3
Resource Requirements Planning

Product line = Pens
Standard labor hours
6-month future planning horizon
Estimate: 1,300 units per month

	Historical			Projected	
Departments	*Units per Month Produced*	*Standard Hours*	*Hours per Unit*	*Standard Hours*	*Per Month (percent)*
Cutting	1,150	4,000	3.478	4,522	24.8%
Secondary	1,150	7,000	6.087	7,913	43.5
Finishing	1,150	2,100	1.826	2,374	13.0
Assembly	1,150	3,000	2.609	3,391	18.6
Total		16,100	14.000	18,200	100.0%

standard hours of work required in each of our four departments (cutting, secondary, finishing, and assembly). If our projected sales were to climb to 1,300 units per month, we could anticipate that the standard hours would increase from 16,100 to 18,200 hours per month. We would use such a report to anticipate the changes we need to make in our headcount or machine count in order to achieve this objective.

There is one additional factor that needs to be considered in the resource requirements plan—the question of whether our production rate is uniform over time. Not only must we consider the seasonality of the demand, but also the potential aggregate change in inventory. The ideal method, of course, is to keep production level over the course of the year, absorbing any fluctuations in the rate of demand with a corresponding change in inventory. This is typical in make-to-stock companies, as shown in the diagram in Figure 11–4, where we can see the principle that the inventory level will be the mirror of the demand when production is uniform.

FIGURE 11–4
Production Planning (level production)

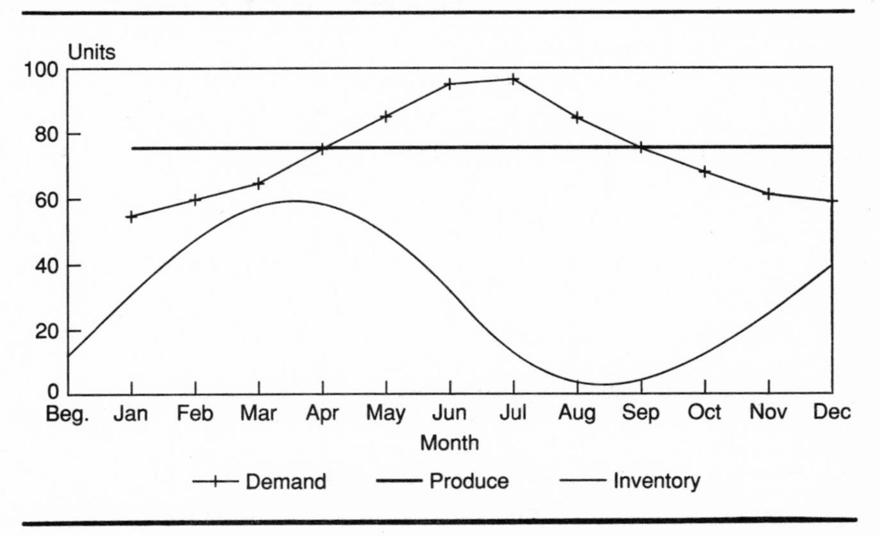

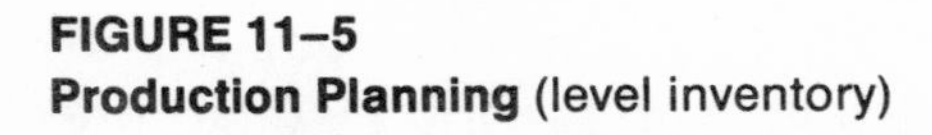

FIGURE 11–5
Production Planning (level inventory)

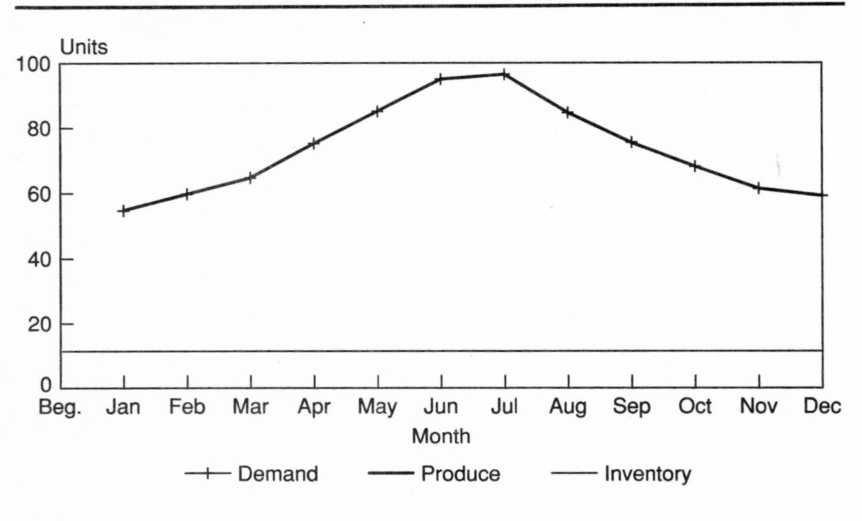

The alternative to this is to be a make-to-order firm, flexing the rate of production to match the demand and thereby keeping the inventory level, as shown in Figure 11–5.

Many companies are unable to change capacity that significantly, in which case they may decide to change the rate of production in a stepped method. In Figure 11–6, we show the strategy of stepping the production rate up once during the year (say, in April), running at a higher rate for three months (until August), and then reducing the output rate to its former level. This causes inventory to move up and down, as reflected in our production plan shown in Figure 11–7.

The resource requirements plan, then, is the place where the overall company strategy—how we plan to produce—is set. There is a veritable law of capacity planning: The overall rate of production is equal to the ending inventory, minus the beginning inventory, plus the expected sales. If we find that, in any period, we cannot flex the production rate to meet the sales rate,

FIGURE 11–6
Production Planning (two-step production)

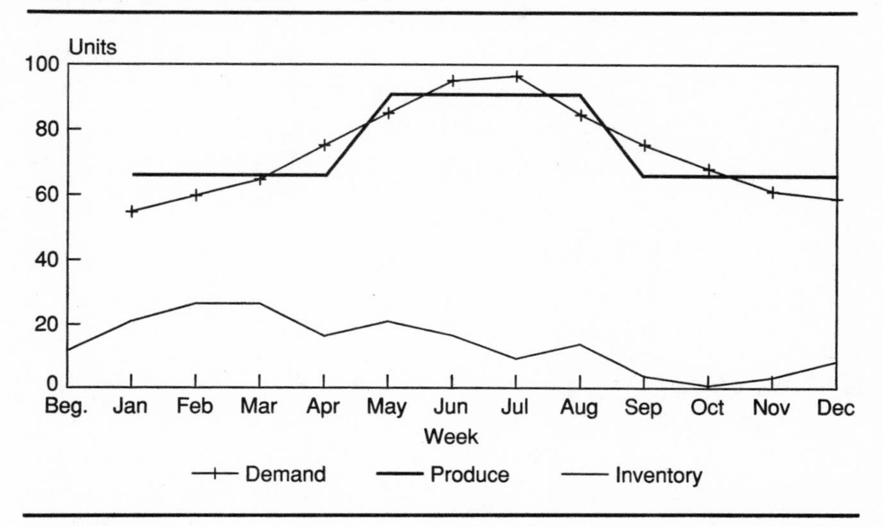

FIGURE 11–7
Production Planning Problem (two-step production)

	Demand	*Produce*	*Inventory*	
Beginning			11	starting 11
January	55	65	21	
February	60	65	26	
March	65	65	26	
April	75	65	16	
May	85	90	21	
June	95	90	16	
July	97	90	9	
August	85	90	14	
September	75	65	4	
October	68	65	1	
November	62	65	4	
December	60	65	9	ending 9
Total	882	880		
Average	73.5	73.3	13.9	

we must build inventory to meet those sales demands. This is sometimes referred to as *stabilization stock.* A make-to-stock company would build it in actual physical inventories, whereas a make-to-order company that wishes to hold its production constant will use a backlog of customer orders to attain this end. The timing of this must be coordinated with the customer in order to keep the backlog from growing and to prevent a negative impact on customer lead time.

ROUGH-CUT CAPACITY REQUIREMENTS PLANNING

Now we go on to rough-cut capacity planning, which takes the master production schedules and multiplies them by the unit product load profile. The result of this calculation gives us the capacity required for key resources (usually people or machines), but may incorporate such things as vendor capabilities or manufacturing floor space and/or warehouse space. Many companies do this only for those work centers considered to be critical work centers or bottlenecks.

Using the example of the master schedule for our pens, we utilize the two-level master scheduling concept, taking the master schedule quantity times our bill of labor to calculate the capacity required in two key work centers, secondaries and assembly. We know that these two areas are critical to the success of our master schedule. We must have enough assembly staff to be able to produce 2,000 pens every period. We know from experience that we typically have shop problems in the secondary operations, so we will create a bill of labor that shows the average time required in each of those two areas. We understand that this is not the actual routing for manufacturing a pen, since (1) the modular bill represents a mix of several different types of pens, and (2) the work of secondary operations is actually performed at a lower level in the bill of material. As we do the master production scheduling at the "Pens" level, we must be sure that it is reasonable to expect that rate of output, as shown in Figure 11–8.

FIGURE 11–8
Rough-Cut Capacity Planning (two-level master scheduling—Level 0)

Level 0

Pens, forecasting bill

	1	2	3	4	5
Forecast	2,000	2,000	2,000	2,000	2,000
orders	100	0	0	0	0
On hand	0	0	0	0	0
Available	1,900	2,000	2,000	2,000	2,000
* MPS *	2,000	2,000	2,000	2,000	2,000

Bill of Labor	Hours per Piece	Rough-Cut Capacity Hours of Work 1	2	3	4	5
Assembly	0.5	1,000	1,000	1,000	1,000	1,000
Secondary	0.1	200	200	200	200	200

In other areas, we may be concerned with such things as the amount of money required for the gold plating needed for that line of pens. At the second level of our master schedule, where we are planning the number of gold upper and lower barrels we need to produce, we could create a bill of labor showing that $1.29 of gold plating for each upper barrel will require $864 in accounts payable funding for Period 1 in order to implement the master schedule. This is depicted in the rough-cut capacity plan shown in Figure 11–9.

If we go through and add up all of the requirements for all of the master scheduled items, the result is a rough-cut capacity plan showing the total amounts of secondary work, assembly work, gold plating, and silver plating, among other things, as shown in Figure 11–10.

The purpose of all of this is to ensure that our master production schedule is a realistic, makable thing. The rough-cut capacity plan is certainly more detailed than the resource requirements plan, since it uses the actual scheduled quantities of items for specific time periods. It is not quite as detailed as the

FIGURE 11–9
Rough-Cut Capacity Planning (two-level master scheduling—Level 1)

Level 1 Explode Available-to-promise

First
25.0% Option gold pens

	1	2	3	4	5
Forecast	475	500	500	500	500
orders	100	0	0	0	0
On hand	95	245	395	545	695
Available	570	650	650	650	650
* MPS *	670	650	650	650	650

		Rough-Cut Capacity Dollars of Plating				
Bill of Labor	$ Per Piece	1	2	3	4	5
Gold plating	$1.29	$864	$839	$839	$839	$839

capacity requirements plan, since we don't show every operation required for every component. It does have the advantages of being relatively simple to create, incorporating a small number of calculations as compared to a full-blown capacity requirements plan. Therefore, it can be easily understood throughout the organization, enabling us to take corrective action for any overloads or underloads that we can see in the future. This

FIGURE 11–10
Rough-Cut Capacity Planning (two-level master scheduling—total amounts)

		Rough-Cut Capacity				
Bill of Labor	Per Piece	1	2	3	4	5
Assembly	0.5	1,000	1,000	1,000	1,000	1,000
Secondary	0.1	200	200	200	200	200
Gold plating	$1.29	$ 864	$ 839	$ 839	$ 839	$ 839
Silver plating	$0.22	$ 286	$ 286	$ 264	$ 264	$ 264

feedback technique is particularly useful in simulating anticipated changes in demand levels that might cause us to consider changing our production rates.

There are, on the other hand, some disadvantages of the rough-cut capacity plan as opposed to the full-blown capacity requirements plan. Inventory quantities are not considered in the mathematics, nor do most companies consider lead time offsets. In our example, this means that the $864 of gold plating in Figure 11–10 might not be needed in its entirety if we have some on hand already, or it may need to arrive in a period prior to the need for the high-level master schedule item. If our periods are days, for instance, we may need the gold material a day prior to the upper barrels going into stock.

Another thing that is not taken into account is that the lot sizes at lower levels may be different from the master schedule lot sizes, thus giving us a different master schedule requirement.

These disadvantages are only troublesome, however, if inventory is to be increased or decreased significantly, or if lead times are very long, or if the majority of the workload consists of a few large customer orders.

Even with these potential shortcomings, rough-cut capacity planning is the real key to success in an MRP system. It is essential that our master production schedule be a doable, workable plan, as described in Focus 11–A. If we exceed the rough-cut capacity, then we must take corrective action to either adjust the capacity or reduce the schedules in order to proceed with execution as planned. In the real world, we may run the

FOCUS 11–A
Focus for Success

Rough-cut CRP must show that the MPS is doable.

master schedule and the rough-cut capacity plan several times to make sure that the production rate creates a supply that will satisfy customer demands.

CAPACITY REQUIREMENTS PLAN

When the master schedule is doable, it will drive the material requirements plan, which explodes through the bills of material, creating the complete component schedules. All of our scheduled receipts plus the planned order releases can be multiplied by the routing for a particular part to calculate the scheduled load for each work center. We have now overcome several of the deficiencies of the rough-cut plan, in that the gross-to-net logic of MRP has netted out our inventory balances. Our lot-sizing technique and lead time offsets should give us a more accurate anticipated quantity and date for each planned work order. This in turn will give us a load for each time period for each work center, as shown in Figure 9–6, reproduced here as Figure 11–11. That load should represent the work that needs to be done in each time period in order to implement all the shop order due dates. As we mentioned in Chapter 9, this is called the infinite loading method, which does not take our available capacity into account. This yields a load profile for each work center that displays the future capacity requirements plan for both planned and released orders. We can then make a comparison of this load to the demonstrated capacity for each work center.

One school of thought maintains that when the required capacity exceeds the demonstrated capacity, it is impractical to expect to produce the orders as scheduled. The term *finite loading* refers to using the computer to rearrange the workload so that any excess load is rescheduled into another available time slot. In Figure 11–12, we indicate that we should move the extra load in Weeks 5/18 and 5/25 out to 6/1. The difficulty in this technique is that we must define some sort of automatic selec-

FIGURE 11–11
Load Profile (infinite loading)

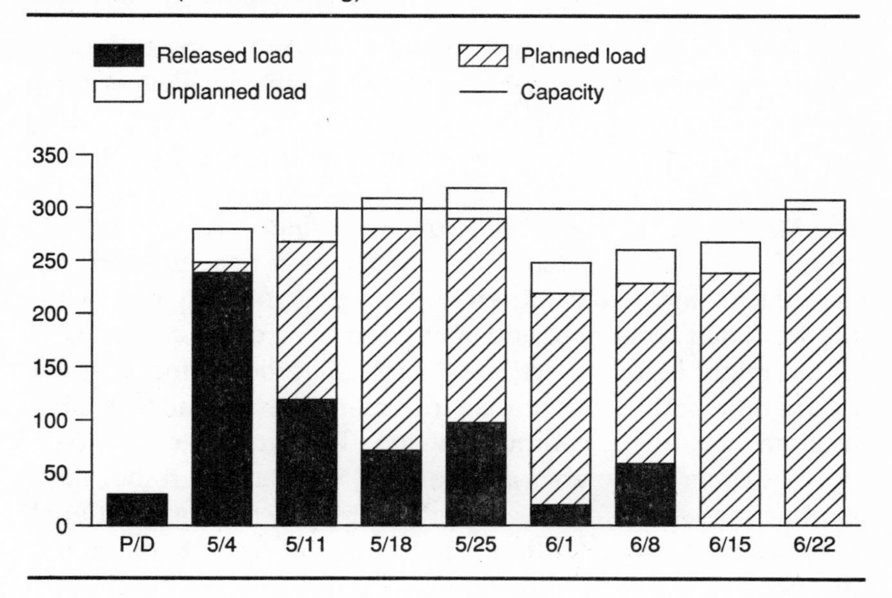

FIGURE 11–12
Load Profile (finite loading)

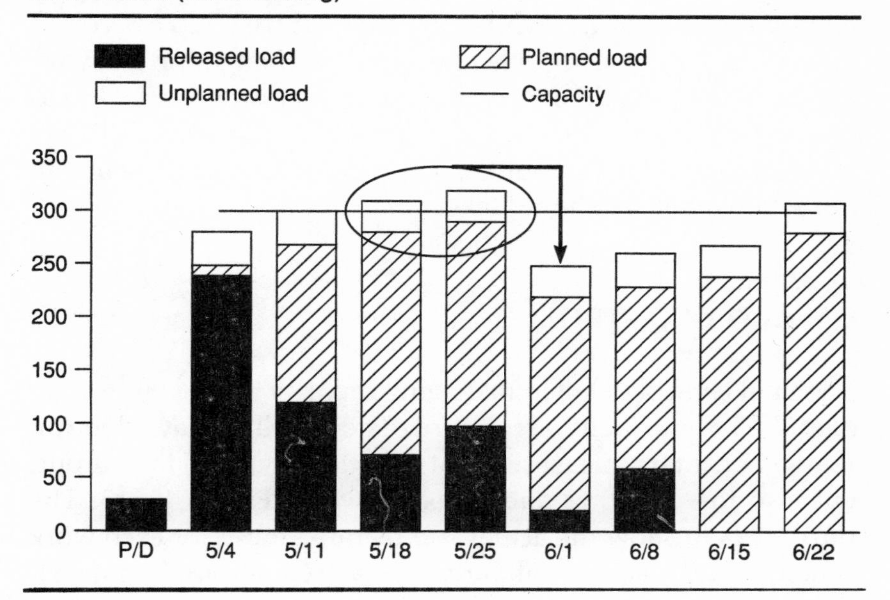

tion process to decide which work orders are going to be delayed. Should it be the last orders loaded, the smallest orders, the largest orders? Or should it be the parts we don't like to make? As you can see, it becomes a complex job to decide automatically which parts ought to be delayed.

Most MRP systems work on the *infinite loading* method, where the work is listed exactly when it is needed, and our manufacturing team can exercise capacity control by reviewing the work that is required and taking some corrective action. This action might be the selection of a specific order to move into a future time period, but human judgment is much more reliable than any automatic computer action. There may, in fact, be a host of other preferable alternatives, such as overtime, subcontracting (which can be thought of as purchasing additional capacity for a period of time), alternate routings, or combining like items to reduce setups.

This capacity requirements planning (CRP) report is usually a very large one, showing many details, so we should be careful to understand that there are a lot of variations that occur in the real manufacturing world; the report provides a very good guideline for the work required, but true success comes from managers who are able to respond to changes on a day-to-day basis. We should expect that the master production schedule and production plans are doable and that the CRP will represent changes in priorities and will become a guideline to the actions that we should take.

CAPACITY CONTROL

The final level of detail in managing capacity is the *input/output control*. This is actually a scorecard of how well we have done on our work centers, measuring not only the input and the output, but also the resulting queue, as shown in Figure 11–13. The purpose is to show the actual and planned input for each work center, and the actual compared to planned output. The devia-

FIGURE 11–13
Input/Output Report

		Period					
		−2	−1	Now	1	2	3
Input	Planned	300	300	300	300	300	300
	Actual	290	300	310			
	Cumulative deviation	−10	−10	0			
Output	Planned	320	320	320	320	300	300
	Actual	330	300	320			
	Cumulative deviation	+10	−10	−10			
Queue	Planned	260	240	220	200	200	200
	Actual 280	240	240	230			

tions tell us how well we are able to adhere to our plan. We can look at the past history of a few periods to determine our demonstrated capacity and make projections into the foreseeable future. We can also measure the queue at a specific point, to be able to compute the anticipated change in that queue. We may see changes that should be made in the work center files; you will remember that the queue is an element of lead time for each work center. We should use the input/output control reports to show us where we should take corrective action in the short term. These corrective actions can include moving workers between work centers, using alternate routings, and scheduling overtime.

Two other popular techniques are *job splitting,* which entails dividing an order into two or more smaller quantities after the order has been released, and *overlapping of successive operations*. This involves moving part of an order to the next work center before the whole lot is finished at the previous work center.

We should not get too carried away with the volume of data available to us on the input/output report. Rather than focus on who did what, we should concentrate on indicators of things that need to be changed, and how we might flow the material through our shop. Many companies moving toward a world-class environment will use simple visual indicators to signal the progress of a job through the shop.

Another technique used in this short-term planning and capacity control is *operations sequencing*. This is a simulation technique that shows the sequence of jobs to be run at each work center based on the current staffing level and load in the short term. This should not be confused with the concept of the operations sequence, a synonym for the routing of a part.

Operations sequencing takes the current dispatch list for the work center in order to perform finite forward scheduling of those loads for a very short period of time. It must include information about the manning level and any anticipated down-time (preventive maintenance, operator vacation, etc.). Given these parameters, it can calculate the queue ahead of time to arrive at a final completion date for each released order. To do this, there must be some tie-breaker logic such as the in-stock due date.

Another tie-breaking method is the *critical ratio*, calculating a priority index number based on the time remaining divided by the work remaining. The lower the ratio, the higher the priority for this job in the queue. Operations sequencing can look ahead to all the work centers to identify bottleneck work centers for management so that corrective action can be taken.

The point of all of this is, simply, that we should manage capacity, our rate of output, by flexing that output to match the input. It doesn't matter what unit of measure we use so long as we are consistent with input and output. With our pencils, we talk about hours worth of work and dollars to the vendor. Other

FOCUS 11–B
Focus for Success

Focus on how we can flex capacity to meet customer demands—not on what the capacity measurement is.

environments can reasonably talk in terms of units, pounds, feet of product, or any other unit of measure. What measurement we use isn't important. We do need to focus on flexibility, as demonstrated in Focus 11–B. We need to be consistent, and most important, we need to be able to take action based on what these measures tell us. We must think of capacity management as opening up the neck of the funnel to a flow of product. All of this information can be summarized in Figure 11–14.

APPLICATIONS

While we show the four main levels of managing capacities, it is not necessary that every company implement all of those levels to the detail indicated. In many world-class companies, it is not

FIGURE 11–14
Open the Bottleneck

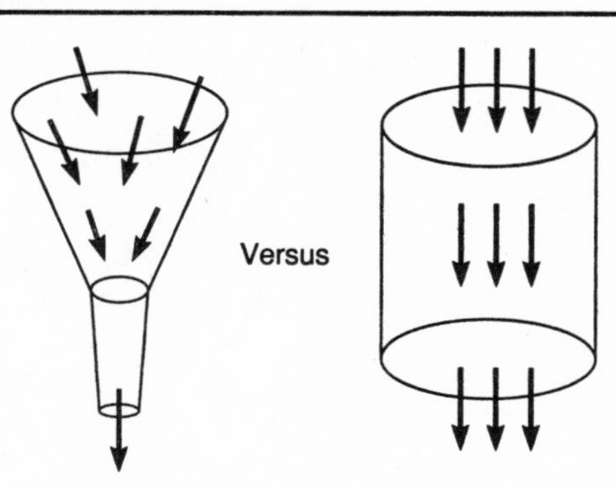

necessary to go to the full level of detail in capacity requirements planning. Process industries dealing with the flow of product down a line may not find it desirable to plan the capacity for each operation, since the master schedules are created indicating flow rate and duration for each product. We can use rough-cut capacity planning to compute the staffing level and duration for each production line. This reduces the amount of detail while providing a perfectly workable tool to advise management of the actions that need to be taken to ensure product supply.

A more discrete just-in-time (JIT) manufacturer would focus on the flow rate of the product, not on the individual work center capacities. We know of one world-class company that has over 15 operations to machine a finished part; but rather than measure capacities and input/output controls for each operation, the operations have been put in a just-in-time cell and capacity is only measured for the cell in total: one work center, one set of data. In the company's production planning, it deals with the flow rate, pieces per day, to ensure that it has sufficient staff to meet the market requirements. The rough-cut capacity only loads the JIT cell in total, and the input/output is a visual control board.

World-class companies make a habit of taking a global view of their capacity problems. A production plan for a highly seasonal item such as Easter-egg color dyes shows all of the demand taking place in about one week out of the year. How can you smooth the production rate for a demand like that? You can increase inventory, or, as one company did, define companion products. The same people and production processes that manufacture Easter-egg dyes now manufacture Halloween face paints. While this doesn't smooth the capacity to a year-round constant, it illustrates the kind of innovative attitude needed to change capacity problems into market opportunities.

CHAPTER 12

MANAGING TECHNOLOGIES

The most important thing to remember about managing technologies is that technologies do not run the company; people do. The vehicle that people use for managing technologies is the assortment of control parameters that we put into the MRP system so that we can manage the way in which the computer makes suggestions to us.

One key control parameter is usually called the *order policy code*. This code has different names in different software systems, but they have the common function of designating which parts are controlled by which logic. For example, the control of an item by MRP logic is different from the control of a master production schedule item. In master production scheduling, all of the planned order releases must be reviewed and accepted by a person before we explode the requirements to a lower level. There are several other order policy codes, such as planning bill items, configured items, and make-to-order items. We will study several examples to ensure that we understand the differences in the logic dictated by these various codes.

The best way to discuss the difference between master schedule items and MRP items is to look at the example of our pen. Let us examine the modular bill for the pens, which we saw last in Chapter 10, when we discussed two-level master scheduling. Figure 10–9 is reproduced as Figure 12–1. Here, you remember, we master scheduled the modular bill called "Pens" as the first level of the master schedule. This part number will have an order policy code to indicate that it is a master scheduled item, and we also want to indicate that it is a planning bill item.

FIGURE 12–1
Modular Bill ("Pens")

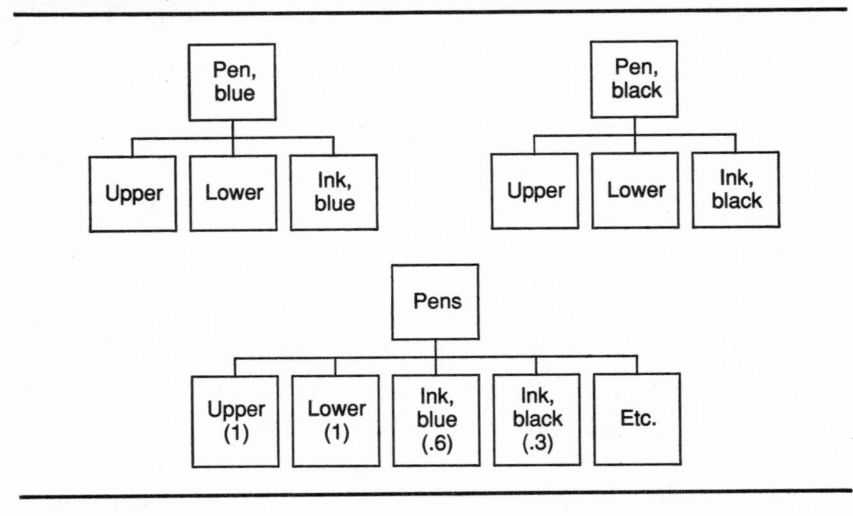

At the second level, we also master scheduled the ink cartridge, so it would have an order policy code to indicate a master scheduled part. We might, for example, call first-level master schedule items Order Policy Code 1 to indicate planning bills, whereas Order Policy Code 2 would indicate a discrete part master schedule item. This is illustrated in Figure 12–2.

If we look at our upper barrel and lower barrel assemblies, we know that the regular MRP gross-to-net logic can be applied to those parts because we will let the master schedule control the requirements for those parts. Consequently, when we master schedule the planning bill, the system will automatically perform the gross-to-net calculations to plan the upper and lower barrels. We will code them, therefore, as MRP items. On the other hand, the ink cartridges also have an independent demand as a forecast for spare parts, and we would be inclined to have a person master schedule those. We don't want to get carried away and master schedule everything; our objective is to master schedule as few items as possible and let the MRP logic take over to ensure that we build matched sets of dependent components.

FIGURE 12–2
Order Policy Code (example)

1	= Planning bill MPS
2	= Discrete part MPS
3–15	= MRP controlled items
40	= Make to order
50	= Configured

You will remember that we configured a customer order for a particular quantity of pens using our planning bill, and we said that the bill of material for that pen would be created for the duration of this order only, so we can call this item a *configured part*, with an order policy code indicating that the item is to be maintained in our database only during the period of time that the customer order exists. In most customer order systems, this policy code indicates that the part number for the configuration is only assigned at order entry and is a temporary part that is automatically deleted, usually after the month-end close for the month in which the customer pays the bill.

One school of thought maintains that we lose historical information about these configured parts. There are two ways in which this technical problem can be overcome. First, we can reconstruct historical purchase information from those parts that we do track. In our example, these configured customer orders consume the forecast for pens, so we can track the sales of pens in aggregate. We can look at the history of sales for various colors of ink cartridges to help us formulate demand assumptions for the various colors of pens. Another method would be to look at the usage of lower barrel assemblies to tell us our total pen sales, gathering data about the demand for individual parts without having to actually track the history of each unique end-unit configuration. It takes some pretty sharp people to be able to accomplish this.

A second method of tracking history is to hold the temporary configured part numbers in the database for some period of time. For example, we might hold all our configured items for six months, after which we would list them all, our people could make evaluations about the product mixes and perhaps make some determinations about parts to keep in stock as a finished good, at which time we would change the part number and order policy codes to an item that would, most likely, be master scheduled. We should be careful about this, because there tends to be a proliferation of parts and configurations, and the database can grow to a point where it becomes quite complex to manage. It is much better to rely on the planning bill method and focus our attention on shortening the lead time to give us better control over the assemble-to-order process.

MAKE TO ORDER

One type of MRP item could appropriately be called a *make-to-order* (MTO) item. These are items that we never want to make unless we actually have a customer order for them. We would code these items with an order policy code (40 in our Figure 12–2) to indicate that we never actually make one of these things unless we have an actual demand. A good example for our pens or pencils would be a specific company logo that would be attached to our upper barrel clip. One of our customers for pens is the organization APICS, which wants its logo to appear on the clip. We don't want to keep these pens in stock until we have a commitment from APICS as to how many pens it wants us to make, so we will code the pens as a make-to-order item.

Other companies may use the MTO order policy for component parts buried deep in the bill of material. One company, for instance, makes steel cylinders with an inner tube and an outer tube. These assemblies may have one of three outer tube diameters, each with the same size inner tube. The normal MRP ordering logic would suggest that we should make and stock the inner tube as a consolidation of the demands for all of the different assemblies. This will give us maximum lot size for this particular component. However, management has determined that the flow of parts is more important than efficiencies of scale. Parts were being manufactured in large quantities, which then sat around while the other tubes were being cut to make an assembly. The company worked instead on methods for minimizing setups to make it easier to change over the cutting lengths. It also coded the inner tube as an MTO order policy code, so that whenever there was an assembly work order, it would create an order to cut both the inner tube and the outer tube to the proper length. These parts were made for the higher-level order only, and did not follow regular MRP logic. More and more world-class companies code their items as MTO items so that they can come out as a consolidated work order to make all components at once and only when there is a higher-level demand.

LOT SIZES

Let us move on to the subject of lot sizing. Within those items that have MRP order policy codes, we can use the order policy code to indicate the lot-size technique we want to use for a given item. We discuss several lot-size techniques in Appendix C. The method that minimizes the inventory is lot-for-lot, which is similar to our make-to-order logic. We want to lean toward using this technique to minimize the batches that tend to lurch through our manufacturing facility. We try to optimize order quantities by using the MRP system, but we should remember the objective of the MRP is to try to bring our lower-level components in to match the master production schedule. Any extra quantities produced consume the lower-level raw materials and productive capacity time. This tends to increase inventories, and therefore costs, without increasing sales. We want to design our MRP with a real focus on the flow of products through our shop in matched sets, as opposed to economic lot sizes. The place for some of the lot-size techniques that weigh the balance between ordering costs and carrying costs is when we are evaluating quantities of very high-priced purchased items.

LEAD TIME

Another major control parameter is the item lead time. We have already discussed the derivation of MRP planned lead times as the sum of run time, plus setup, plus queue, plus all other interoperation times. One control parameter in the MRP system needs to be the queue time in individual work centers, which is frequently maintained in the work center file. This is where we put our estimate of how much time an order will wait in line before it actually gets dispatched to a work center. This number is typically used in conjunction with the routings to calculate the MRP planned lead time. Human beings have a tendency to think in terms of increasing those numbers to allow orders more time to be work in process. This is a very dangerous practice and can cause lead times to go into an inflationary spiral that can literally overload a shop. For example, if our queues are large, manufacturing will say that we should increase the lead time for components to give us more time to get them through a large queue. If we succumb to the temptation to increase the queues in the work center file, the MRP lead time will increase. When the lead time increases, planned order releases will be released to the shop earlier than they would have been otherwise. This causes more work to go out in the shop, and so the queues will grow longer. In *Production and Inventory Control: Applications*, George Plossl calls this the "vicious manufacturing cycle."[1] The secret to breaking this cycle of queue increase/lead time increase/queue increase is in the hands of the people. There are two things that can be done to stop the inflation. First, we should increase the capacity in the short term for those areas that have large queues, in order to push the bubble on through and out of the shop. Then, we should look at the flow of work through the shop with the object of reducing the size of queues

---

[1] Published by George Plossl Educational Services, Inc., Atlanta, Ga., © 1983.

so that we can increase the speed of throughput within the shop.

The other way is not to calculate MRP planned lead times using all the detailed lead time data, but rather to go to the lead time field in the item master and *make it smaller*. This takes an integrated team effort between the planning functions and the manufacturing functions in order to ensure that the MRP planned lead times truly reflect the time it takes to flow through our shop and, at the same time, be as short as possible.

A companion concept is the actual lead time. This is usually not an actual control parameter tracked in the MRP system, but in world-class companies today, it is being more frequently tracked in actual orders to give us a benchmark of our ability to perform. The objective is that, wherever we are now, we should strive to reduce the lead time even further. It is important that we understand the principles that determine the actual lead times. There are two: (1) size of the queue and (2) priority of the order. The larger the queue, the longer it's going to take the order to go through and the longer the actual lead time is going to be. This is getting to be a familiar concept. Learn it by heart. If an order has a higher priority—that is to say, a closer due date—it will get worked on first and therefore have a shorter actual lead time. Real success comes to those companies that keep their actual lead time short by keeping their queues low and the time between the planned order release date and the due date short. In summary, we should focus on keeping the queues low, the lead times short, and the lot sizes small, as shown in Focus 12–A.

Another consideration is purchased parts. Here, we may not have as much control over the lot sizes or the queues, but the same principles apply. We want to consider the supplier as an extension of our own facility who just happens to have machines and workers at a site removed from our own. Just as we need teamwork among our own people, we want to have a good relationship with our vendors. There is sometimes a tendency for a vendor's salesperson to tell us that the vendor's

FOCUS 12–A
Focus for Success

Focus on keeping:
- Queues low
- Lead times short
- Lot size small

backlog is growing, and that we, therefore, should increase the lead times for our components. If we do increase the lead time for those components, it will cause more planned orders to be released, causing the vendor's backlog to grow. We are caught in the lead time inflation cycle again. We should be more open about our future planned order releases, enabling the supplier to anticipate the capacity that will be required to fill our needs. This is sometimes done through a formal blanket order, or a purchase order showing future demands. For example, a purchase order might be broken up into three windows of time: All the requirements for the next three weeks can be fabricated and delivered. For six weeks beyond that, the supplier is authorized to purchase raw materials to satisfy those requirements. For all requirements beyond the nine weeks, the order allows only advance capacity planning. Other companies might be less formal about this scenario, merely scheduling meetings with the supplier at specified frequencies. All of this could lead us to the concept of using the MRP only as a signal system to our vendors, where the actual authorization to fabricate and deliver is based on some type of visual pull system.

Another major consideration is the flow of work through the shop. Our dispatch list shows the order priorities for the work flowing through our work centers. The key date on the list should be the date that the part is due in stock. The people who are deciding the sequence of orders to go through our work centers should utilize several tools to ensure that the parts get into stock when they are due. Therefore, a dispatcher should not look at just one operation. When making decisions, the dispatcher needs to look at the other work centers and evaluate the operator utilization and the flow of this product. In some instances, it makes a lot more sense to *not* work on a part if a downstream machine is bottlenecked so severely that the parts won't get through. The real solution is to increase the capacity at the bottleneck work center. One effective way of doing this is to release some labor from one work center to go to the bottleneck.

Other factors that come into play are *horizontal* and *vertical dependencies*. What we mean by horizontal dependencies is that our particular component should flow through the shop in coordination with other components that go into the assembly. For our pencil, we have an example of this in Figure 12–3, where the eraser assembly is made of two components, the eraser and the eraser sleeve. If you are the dispatcher for the operation to form the eraser sleeve, you are most interested in the status of the erasers that are going to come together with the sleeve to make the assembly. In a formal MRP system, we use the concept of

FIGURE 12–3
Eraser Subassembly

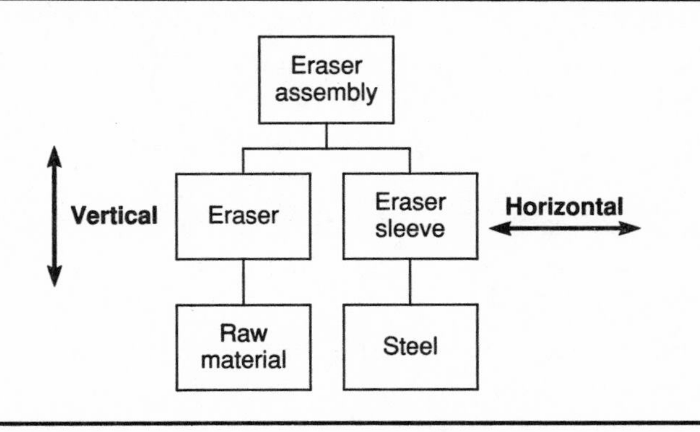

pegging to find the part number of the eraser assembly that is required, and then to find the other components that are needed in addition to the eraser sleeve in order to build the assembly. This is a horizontal dependency. The best of all worlds is to time the flow of the two components, manufacturing them together rather than as individual batches. Of course, an alternative that world-class companies are using today is to lay out work centers into cells so that the operators can use visual methods to coordinate the progress of the two components. Here is where we talk about the minute-to-minute execution of the planning system out on the shop floor. The MRP planning logic is effective in scheduling the receipts of raw materials; however, just-in-time visual systems are more effective in coordinating the flow of production through our shop. We shouldn't count on the MRP system to be terribly effective in the quickly changing shop floor environment. MRP is most effective at the cumulative lead time of the component.

BILLS OF MATERIAL

Now we turn to the technique known as *flattening the bill of material*. This is the process of looking at our bill of material tree and also looking at the process we go through in the shop to actually build those parts. Following the MRP logic in its purest form, we would create a component and its corresponding MRP plan for each level in the bill of material, as shown in the example of the pencil, Figure 12–4. This implies that we will create subassemblies for upper barrel, lower barrel, eraser, and the four spare leads. Traditional MRP logic demands that we put those four components into stock before we can use them to build a pencil assembly. In fact, we can be confident that the only time we make lower subassemblies is when we have a requirement for that specific pencil. This is referred to as a vertical dependency. We might consider coding the lower barrel as a make-to-order item, as we discussed above, or as a phan-

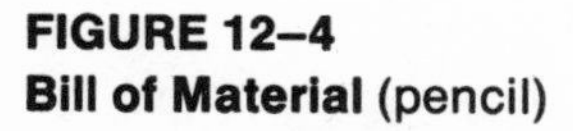

FIGURE 12–4
Bill of Material (pencil)

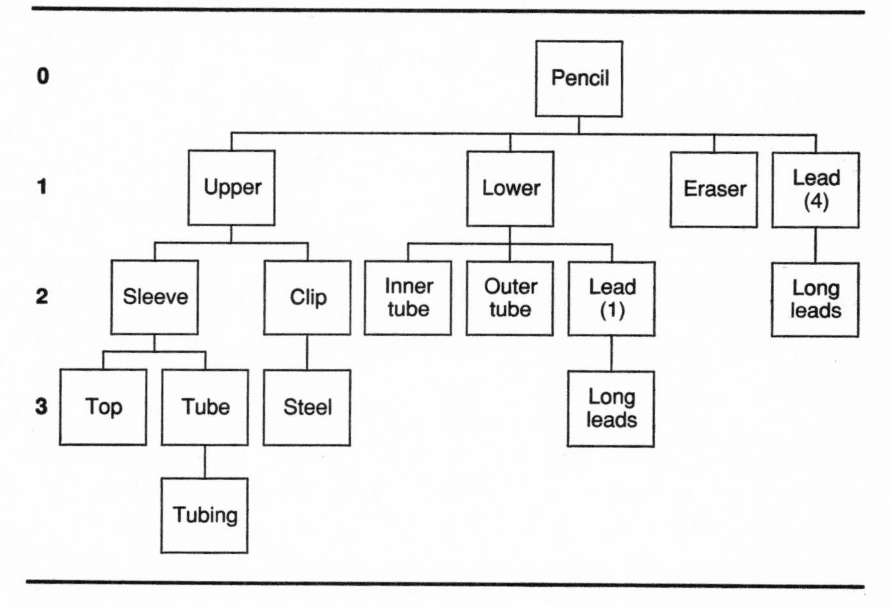

tom. Either method will have the effect of flattening the bill for that one item. However, if we look out on the shop floor, we can see that we have an assembly line that builds the pencils complete from their several components. It makes no sense to actually build a lower subassembly, process the work order paperwork, put the assembly into stock, and then issue it back out again in order to make the pencil. What we want to do is flatten the bill for the pencil so that the order calls for the components that make up the subassemblies. Looking at Figure 12–4, we can see that three of the items on Level 1 will only exist on the pencil assembly line. It would make sense to stock the leads cut to length because they are used at several levels in this bill as well as for spare parts. It makes sense to omit the various subassemblies and to plan and schedule the components to arrive at the proper places on the assembly line. Those components would be the top, clip, upper barrel tube, leads, lower inner tube mechanism, lower outer tube barrel, point, eraser, and eraser sleeve. Every time we make an order for pencils, we want the MRP system to plan those components to be delivered straight to the point of use on the line. It might even look something like Figure 12–5. The routing to manufacture must include the operations required to assemble the pencil, plus the operations required to make the subassemblies.

Notice that we now need to process the transactions for all the various components as they are delivered to our assembly line. There is a better way to handle all these transactions, known as *backflushing*.

Looking at the pencil, we can say, for example, that the eraser assembly can be used in both the gold and silver pencils. The leads would be common to multiple pencil types. We can set up an assembly line where the components are stored at the line and readily available to the operators for whatever assembly they are making at the time. Rather than process transactions for each component, we should issue parts from the stockroom to the assembly area in large quantities, as shown in Figure 12–6.

Let us consider erasers as an example. We move erasers in a convenient quantity, and we move eraser sleeves as well. If we decide to move 5,000 erasers at a time and deliver them to the assembly line, we would process a transaction moving that

FIGURE 12–5
Pencil Assembly (U-line)

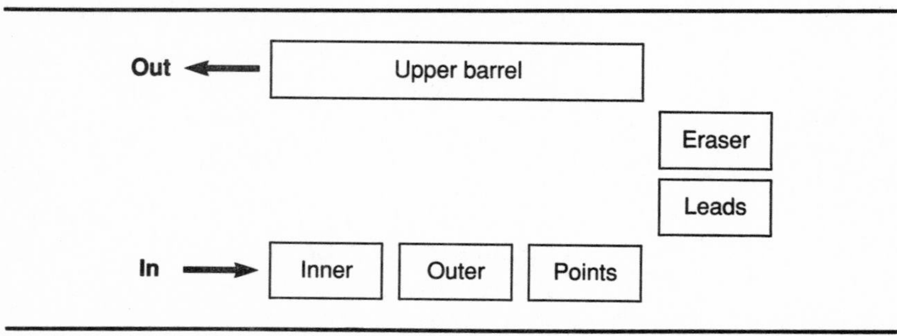

FIGURE 12–6
Backflush Method (pencil assembly line)

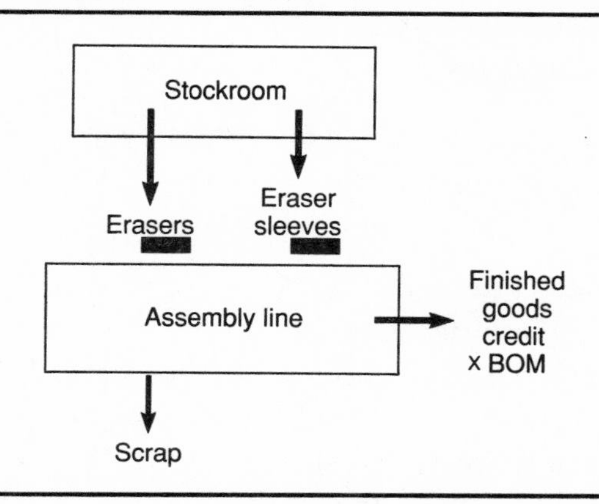

quantity from stockroom inventory to a work-in-process location. At the end of the assembly line, we would process transactions showing the quantity of pencils produced. If we produce, say, 3,200 pencils today, we can look up the bill of material and determine that we must have used 3,200 erasers and 3,200 eraser sleeves. We can now reduce the balance on hand at our work-in-process location from 5,000 erasers to 1,800 erasers. We can also deduct the 3,200 eraser sleeves. This is known as backflushing to relieve work-in-process inventory.

The main advantage of this technique is that we have minimized the number of transactions and work orders that we must process through our MRP system. In most industries, the work orders such as those for assembled pencils would be for very small numbers of many different styles every day. Doing this using the classical MRP approach would create a work order for each eraser used in each pencil and process transactions for each of the components. We can tell by looking at our pencil that we would have the same situation with the upper barrel assembly and the lower barrel assembly. For many companies, this creates a huge number of work orders and transactions, and the company can become bogged down in paperwork. By flattening

FOCUS 12–B
Focus for Success

Focus on the flow of parts, not MRP level-by-level processing.

the bill and using the computer to calculate quantities of parts that must have been used, much of this unnecessary detail is eliminated. As Focus 12–B points out, we need to concentrate on the flow of material in our manufacturing process.

The disadvantage of backflushing, on the other hand, is that we need some way of accounting for scrap that will reduce the balance on hand of our components. We also need to take an inventory of these work-in-process locations every so often so that we can detect any variances that might occur.

Once we have control of the backflush method, however, we will see a significant reduction in the amount of paperwork and the number of items that must be processed. Even in our example, the building of a pencil assembly requires only one work order, whereas it would have required four work orders had we used classic methodology.

KEEP IT SIMPLE BUT ACCURATE

One thing that hurts many companies is that they implement the MRP system using very strict rules stating that they must process every transaction for every level in the bill of material. It is much better for a company to look very closely at their bills of material, planning, and shop processes in order to flatten them as much as they possibly can. In our example, the eraser might be cut to length at a machine right next to the assembly line. Then we could draw long pieces of raw eraser stock from our vendor and deliver them directly to the cutting machine in its position next to the assembly line. This would achieve the best that an MRP system can give. We use the master schedule for pencils, apply the MRP logic to tell our vendor that the raw eraser material is needed and when we need it, and let our backflushing issue the eraser material for whatever pencils we make. These concepts can be extended across our whole array of components.

To do these things, we must focus on a high degree of data integrity. In order for us to utilize a flat bill of material, the bill of material must be absolutely accurate—100 percent of the items we manufacture must have the correct bill of material. Historically, there have been long debates over which bill of material—engineering, accounting, manufacturing, or planning—to use as the "right" one. If our company is to supply our customers with product, there is only one answer; the correct bill is the one that tells us how the product is actually put together. All of our planning systems must accurately reflect what's happening out on the shop floor.

Another issue of data integrity is inventory record accuracy. Since the MRP uses balance on hand as the basis for planning each individual part, it is important that this balance on hand be extremely accurate—98 percent or higher. If we have not achieved this level, our focus should be on processing our transactions with a view toward uncovering and eliminating any cause of error. For our MRP to be a success, we must have in place a cycle-counting system that audits our accuracy. While we may have some items that are controlled using high tolerances and that are not worth the trouble it takes to track them closely, aggregate inventory accuracy must remain high.

A third area of data integrity should be schedule adherence. We should have the objective of at least 95 percent schedule adherence at the master schedule level. This means that of items that are ordered with a due date, 95 percent will be completed on that due date. The objective is to manage the process of building the parts. We don't want to be late, which will hurt customer satisfaction, but neither do we want to be

FOCUS 12–C
Focus for Success

Focus on *continuous improvement* by *identifying* and *eliminating* causes of errors and/or delays.

early. That would mean that our planned lead time is not accurate and that we have excess inventory. As in cycle counting, we should use our schedule adherence report to help us identify and eliminate the causes of scheduling problems.

There are, in some companies, many other technical issues to be resolved. These companies may even use other parameters to control the MRP system. What we have discussed here are the major concepts that need to be taken into account. It is important to measure where we are today and what steps we need to take to help us move forward. We should focus on the control parameters that we need to adjust and the steps that we need to take to stay on the road to continuous improvement. This is done most effectively by identifying the causes of any problems or delays and taking steps to eliminate those causes. Then we should enter those new control parameters into our MRP system and start the process all over again, as shown in Focus 12–C.

APPLICATIONS

World-class companies take these concepts and apply them in many creative ways. For example, one plastics extrusion company using an MRP system has designed its system to fit the manufacturing process. One thing it did was consolidate its setup process sheets into a "setup book" kept on the shop floor. The company doesn't try to keep all the data relevant to plastic extrusion setups resident in its MRP computer system. The MRP system also enters orders for customer parts that have the same base materials as one work order, independent of length. One or more part numbers of differing lengths, but using the same material compounds, will be produced as though they were one work order. This is done by creating a bill of material that has several levels of phantoms that convert multiple lengths of end unit into a common base material. In the MRP system, this creates one work order for multiple parts, reducing paperwork and producing all customer requirements off one setup.

Let's consider backflushing again. One high-volume repetitive manufacturing company achieved major improvements in its manufacturing process when it was able to carry the backflushing process all the way back to the raw material level. A steel fabricated part required 8 different manufacturing work orders, composed of 12 different purchased parts. The bill of material was four levels deep, requiring 19 different transactions to build a completed unit. After reviewing the process and forming a shop assembly line, the company was able to flatten the bill of material to two levels. Now it could bring the 12 purchased parts together with only three manufacturing work orders and a total of four transactions. Today, the company is still looking for ways to reduce the three orders to one. It was able to achieve a significant increase in inventory record accuracy and order variances were virtually eliminated. The backflush routine, after two years of problem solving, is now down to less than .1 percent variance. Perhaps the most significant customer advantage in this improvement was the overall lead time improvement for the end-unit product, which dropped from 14 weeks to 4.

In some companies, it becomes very expensive to rearrange the shop layout into assembly work cells. One muffler manufacturer with many specialized pieces of machinery that fabricate muffler components found that not all these machines could be moved close to the assembly line. The company was able to flatten its bill of material significantly and to link its operations together by using overhead conveyers. If there is an empty hook, you can build the part and put it on the hook; a series of hooks is grouped together on the conveyer in such a fashion that when all the hooks are filled, the parts are available to build an assembly. The hooks have become, in effect, a visual signaling device. The only dispatch list required is for the end-unit assemblies. The MRP becomes the tool to plan the raw materials coming into the plant and just-in-time visual signals become the execution tool.

SECTION 5

PEOPLE

CHAPTER 13

NET CHANGE MRP SYSTEMS

People are the most important part of a successful MRP system. In all our previous discussions, we talked about the techniques we use to create information that gives us the ability to react quickly to changes. There are several techniques that have been used to successfully manage these changes. However, before we describe any of these techniques, it is necessary for us to remember that the most important aspect of our MRP system is what people do with the information that is generated. People must assimilate and make use of the suggestions. It does no good to make suggested changes if we are unable to review those changes and take effective action.

There are at least three things that people must react to (see Figure 13–1). The first two, planned order releases and reschedule notices, appear on our exception reports, as we discussed in Chapters 7 and 8. The third, *capacity overloads,* occurs in our capacity planning report when we have too much work to be done in any specific time period. We should also be concerned with those manufacturing operations when we have more time than we have work. The ultimate solution is to move people to the overloaded work centers so that we can build all of the parts required by the master schedule. If we have a machine capacity problem, we need to take more complicated remedies, such as alternate routings, subcontracting, or as a last resort, changing the requirements. The evaluation of these various alternatives takes time and effort.

The real value of the MRP system is that it generates the data that shows where we stand today. Given the master schedule and the requirements for all the component parts, it shows

FIGURE 13–1
MRP Suggestions

- Planned order releases
- Reschedule notices
- Capacity overloads

the new orders that must be released, plus any reschedules, plus any capacity requirements. We—people—need to take actions based on this data. As we implement these actions, we tell the system what actions we took. This takes the form of releasing orders, changing due dates on existing orders, and adjusting the capacity in our manufacturing shop. What we expect is that the next run of our MRP will tell us the results of these adjustments. The real key to the MRP system is this iterative process: The system suggests to us what actions need to be taken, we take those actions, recompute, and the system shows us the results. The way people use this dynamic information source is very important.

There are two main ways of manipulating this data: *regeneration MRP* and *net change MRP*.

REGENERATION MRP

Simply put, regeneration MRP is the process of taking the master production schedule and exploding it down through the whole bill of material. From this process we will totally regenerate the gross requirements, which will lead in turn to regenerated planned order releases and perhaps reschedule notices.

The common way to look at this, as we have been doing all along, is in weekly increments. We can think of it as shown in Figure 13–2 (which we first saw as Figure 3–11 and later as Figure 8–1). If we regenerate this whole worksheet once a week, we can reestablish the priorities for all the components needed to build the master schedule.

FIGURE 13–2
MRP Worksheet (upper barrel)

	Now	Period 1	2	3	4	5	6	7	8	9
Requirement		20		20	20		5		20	
Scheduled receipts			25							
On hand	25	5	30	10	15	15	10	10	15	15
Planned order releases		25				25				

Suggest reschedule: 25 from Period 2 to due Period 3

Let's take a look at our time buckets. If each of the columns in the worksheet represents a week, and we regenerate the MRP every week, we are doing what is commonly called "rolling the MRP ahead a week." In the figure, Week 1 is now past due and a new period, Week 10, would roll onto the right-hand side of the chart. What this does is establish that our dates and priorities are no more precise than the size of the time bucket. If the bucket is a week long, the worksheet won't tell us if the action is due on Monday or Friday of that week. We need to establish a convention in our minds to tell ourselves that all the events occurring in a time bucket happen all at once. We might want to say that everything magically happens at once on Wednesdays at noon. In the real world, however, this is not the case. We know of one plant, for example, that always schedules customer shipments on Friday and schedules the work orders on Monday. Our MRP worksheet gives us priorities no more accurate or refined than "the week of [date]."

Many situations require priority planning much more refined and accurate than "the week of . . . ," so we can change the size of our time bucket to represent a day rather than a week. Then our MRP worksheet might show each time bucket representing a particular workday. Looking at our worksheet, we might call Periods 1 through 5 Monday through Friday, so our requirements will be maintained in daily increments. If we regenerate the full explosion process once a week—say, Monday morning—it would show accurate priorities for that point in time. But as time passes and we move toward Friday, we may discover that our priorities have changed over the course of the week. This leads us to the conclusion that our priorities will not be any more accurate than the size of our bucket or the regeneration period, *whichever is larger.*

Many companies have begun to regenerate the MRP on a daily basis in an attempt to deal with this reality. One of the main advantages is that valid priorities are maintained for each day. As customer requirements change or we make changes in schedules or capacity, we can see the effect of these changes on a daily basis. Another advantage is that we can see what work orders should be released each day, thereby smoothing the flow of work into the shop.

There are some disadvantages to this daily regeneration, however. One of the main problems is the proliferation of data that we expect people to assimilate and act on. Remember to be wary of Plossl's "illusion of precision." MRP is most effective at the lead time of components, and therefore, is best used to plan raw material flow from vendors. As we need smaller and smaller time buckets, we should investigate other just-in-time visual techniques to control the flow of work through our shop. Another disadvantage is the question of whether we have sufficient computer power to handle this massive data processing task. In a company where it takes six hours to run the MRP, the computer would be tied up for the greater part of every day. A company with higher-powered computer capabilities might be able to run its MRP in less than an hour, and for them, daily regeneration is a practical step.

Another disadvantage to daily regeneration is that a short time bucket such as one day might mean that we need many buckets to get us to our lead time. If lead time for a particular component is six weeks, we might not have our requirements refined enough to tell us whether we need delivery on Tuesday or on Wednesday six weeks from now. As we get further into the future, we might want to change our bucket size from days back to weeks, conserving report-running time. This is known as the *"variable bucket"* method.

What is really important in all of this is that the bucket size is a convention that we assign to help us print our reports. It is really best to think of the MRP as a "bucketless" system, where all the data is stored in association with a specific date. The date represents the *date needed* for a requirement, or the *date due* for a supply. In any system, all the data is stored this way, anyway. We only use the convention of a specific bucket size on our worksheet as a visualization aid. All of our suggestion notices can be printed with specific dates attached.

There is a wide spectrum of industries with varying requirements, but few of them can operate on a planning scheme that defines priorities merely as "the week of . . . . " Most companies need, at a minimum, to be accurate to the day. So, to maintain our priorities, at least in the short term, we need to operate in the mode of daily regeneration. If we have insufficient computer capability, there is an alternative available to us in *net change MRP*.

NET CHANGE MRP

Net change is commonly referred to as *transaction-triggered MRP*. A change in the requirements, orders, inventory status, bill of material, or anything else that might affect the item could precipitate a change in the components for that item. We would do a partial explosion of those requirements, and hence, the MRP process is continuously operating within our computer.

There are some distinct advantages of this method over the regeneration method, as shown in Figure 13–3. One advantage

FIGURE 13–3
Net Change MRP

Advantages
1. Transaction triggered
2. Continuous updating
3. Quick reaction time

Disadvantages
1. Requires strict disciplines
2. Must control nervousness and chatter
3. Data processing efficiency

is that any transaction to a part will show the effect that transaction has on the MRP plans for the supply and demand of that part. We are continuously updating these plans and can expect a quick reaction time from the system. This quick reaction time is a mixed blessing, as it points toward several disadvantages. First, we must have very strict disciplines in place. Our data entry and data integrity verification procedures must be extremely reliable. We must also be able to control the "nervousness" or "chatter." This is caused when we implement a transaction, see that its results are impractical, and delete it immediately. This causes our net change to switch back and forth rapidly. This not only confuses people using the system, it can be the cause of inefficiency in the data processing arena. What we really have is an MRP that is constantly running with these partial explosions; a "ripple effect" is taking place continuously as one element of the data changes and has an impact on related items.

An example of how this works is shown in Figure 13–4. You can see the requirements for the eraser assembly show a planned order release of 95 in Period 1 and 70 in Period 4. Of course, planned order releases at one level create gross requirements at the next level, so the gross requirement for eraser sleeves will be 95 in Period 1 and 70 in Period 4. With a balance on hand of 100 eraser sleeves, you can see that the planned

FIGURE 13–4
MRP Worksheet (net change MRP)

Eraser assembly

Lead time = 3
Lot = 3 periods

	Now	1 5/21	2 5/28	3 6/3	4 6/10	5 6/17	6 6/24	7 7/1	8 7/8	9 7/15	10 7/22
Requirement		25	20	35	50	25	20	30	20	20	30
Scheduled receipts											
On hand	80	55	35	0	45	20	0	40	20	0	0
Planned order releases		95			70			30			

Planned order for 95 due 6/10 Release now
Planned order for 70 due 7/1 Release 6/10
Planned order for 30 due 7/22 Release 7/1

Eraser sleeve

Lead time = 3
Lot = Lot-4-Lot

	Now	1 5/21	2 5/28	3 6/3	4 6/10	5 6/17	6 6/24	7 7/1	8 7/8	9 7/15	10 7/22
Requirement		95			70			30			
Scheduled receipts											
On hand	100	5	5	5	0	0	0	0	0	0	0
Planned order releases		65			30						

New order for 65 due 6/10 Release now
New order for 30 due 7/1 Release 6/10

order release to the vendor for 65 eraser sleeves must be released in Period 1 to come in at Period 4.

A net change system requires very strict disciplines. Suppose that our balance on hand is inaccurate. Our cycle counters report that, instead of 100 eraser sleeves, we only have 50 in the stockroom. We implement this change, which will give us the situation in Figure 13–5. Here, we see that we need to place an order with the vendor for 70 due in Period 4 and an emergency order of 45 for yesterday. On the other hand, if the cycle counter comes back half an hour later and says that the other box of 50 was just found, we call the vendor up to cancel that emergency order altogether and set the other order back to 65 from 70. The printed reports that we get after all of this has happened would cause a certain mistrust in what the computer was telling us, and people would tend to think the computer was "crying wolf" and shouldn't really be believed. This illustrates the sort of strict discipline we need to adhere to when we tell the system of data changes. Unless we're sure of our facts, we should not input changes in bills of material or cycle count variations or the myriad of other changes that come up in the course of business.

There are other transactions that, by their very nature, cause a slight change in the MRP logic. One of these is called *allocation*. In an MRP system, a component item is allocated when an order to pick the part has been released but the part itself has not been issued from our stockroom. This is shown in Figure 13–6, where we have converted the planned order release for 95 eraser assemblies into a scheduled receipt due in Period 4. When we tell the system we have converted a planned order release into a scheduled receipt, we generate a pick list for the stockroom to pull the components. At that point, the eraser assembly MRP worksheet would remove the planned order release in Period 1, as shown in Figure 13–6. It takes a little time, though, before our stockroom people actually reduce the balance on hand by picking those 95 eraser sleeves. We still have 100 sleeves on the stockroom shelves, and 95 are needed for the new order in process. It would ordinarily be shown as an allo-

FIGURE 13–5
MRP Worksheet (net change MRP—chatter)

Eraser assembly

Lead time = 3
Lot = 3 periods

Period	Now	1 5/21	2 5/28	3 6/3	4 6/10	5 6/17	6 6/24	7 7/1	8 7/8	9 7/15	10 7/22
Requirement		25	20	35	50	25	20	30	20	20	30
Scheduled receipts											
On hand	80	55	35	0	45	20	0	40	20	0	0
Planned order releases		95			70			30			

Planned order for 95 due 6/10 Release now
Planned order for 70 due 7/1 Release 6/10
Planned order for 30 due 7/22 Release 7/1

Eraser sleeve

Lead time = 3
Lot = Lot-4-Lot

Period	Now	1 5/21	2 5/28	3 6/3	4 6/10	5 6/17	6 6/24	7 7/1	8 7/8	9 7/15	10 7/22
Requirement		95			70			30			
Scheduled receipts											
On hand	50	0	0	0	0	0	0	0	0	0	0
Planned order releases	45	70			30						

Emergency past due order 45 due 5/21
New order for 70 due 6/10 Release now
New order for 30 due 7/1 Release 6/10

FIGURE 13–6
MRP Worksheet (net change MRP—allocation)

Eraser assembly

Lead time = 3
Lot = 3 periods

	Now	1 5/21	2 5/28	3 6/3	4 6/10	5 6/17	6 6/24	7 7/1	8 7/8	9 7/15	10 7/22
Requirement		25	20	35	50	25	20	30	20	20	30
Scheduled receipts					95						
On hand	80	55	35	0	45	20	0	40	20	0	0
Planned order releases					70			30			

Scheduled order 38909 for 95 due 6/10
Planned order for 70 due 7/1 Release 6/10
Planned order for 30 due 7/22 Release 7/1

Allocation

Eraser sleeve

Lead time = 3
Lot = Lot-4-Lot

	Now	1 5/21	2 5/28	3 6/3	4 6/10	5 6/17	6 6/24	7 7/1	8 7/8	9 7/15	10 7/22
Requirement	95				70			30			
Scheduled receipts											
On hand	100	5	5	5	0	0	0	0	0	0	0
Planned order releases		65			30						

New order for 65 due 6/10 Release now
New order for 30 due 7/1 Release 6/10

cation quantity on the MRP worksheet, represented in most cases as a gross requirement prior to Period 1. The quantity would remain in the allocation field until the time when the stockroom actually picks the parts, when the quantity picked would be subtracted from both the balance and the allocation field. This timing disparity between the MRP report and the human activity in the stockroom means that we must have allocation logic as a part of any net change MRP system.

The allocation requirement and the stricter disciplines are both necessary when running a regeneration MRP on a daily basis. If we don't utilize the allocation concept, any planned order release to be converted to a scheduled receipt must be picked from the stockroom on the same day it is released. While this might be the way things usually happen in the plant, it is dangerous to assume that it *always* happens this way.

Allocations have another use. You will recall that in Chapter 6, we discussed the what-if capability of the master production schedule. When we simulate a prospective master schedule, we must be careful not to steal parts that have been allocated to customer orders. One means of guarding against this is to subtract the allocated quantity from the current balance on hand in order to run the what-if using available balance only. This available balance can be used to make determinations of true shortages on specific customer orders.

APPLICATIONS

There are all sorts of ways that companies implement these processes. Perhaps the most common one is exemplified by a hydraulic motor manufacturer we know who regenerates the MRP routinely every night. The main purpose is to print the dispatch lists at the beginning of the first shift so that priorities can be maintained on a daily basis.

Sometimes daily is not a refined enough time bucket. A just-in-time manufacturer supplying automotive assembly plants needs to know the priorities within a four-hour window —half a shift. It doesn't even run its MRP with conventional dates—it uses indicators for date, shift, and which half of the shift is involved. For example, 06/14/YY/A/1 would represent the first half of Shift A (YY, of course, represents the year). In effect, the company is maintaining its priorities in a four-hour bucket. We would advocate that in such a case, the MRP should be used only as a supplier forecasting system and that assembly should rely on visual signals, but this company was effective in its use of the MRP system as a vehicle to move the corporate culture toward a more stringent scheduling system.

CHAPTER 14

IMPLEMENTATION

"If we only had management commitment, this thing would really fly." How often have we heard that? And exactly what do we mean when we talk about management commitment? In our opinion, there are two elements to this elusive phenomenon: backing and ongoing involvement.

MANAGEMENT BACKING

Backing includes the usual assortment of actions that management takes in deciding to install a new manufacturing system. In the case of an MRP system, this involves the actual writing of the check for the hardware and software. Normally, that check is pretty big, too, which means that approval for the expenditure may not come easy. The first step in securing this management backing is, therefore, a first-cut, high-level education overview of the benefits and costs associated with a really first-class MRP system. At this point, we should begin to crystalize specific objectives that management expects the new system to achieve. Specific objectives are *not* things like "improve customer service," or "manage inventory more effectively." Like motherhood and apple pie, these are emotionally positive concepts that everyone feels warm and fuzzy about. We need to go through a process of formally identifying tangible goals that have elements of measurement and time associated with them. "Improve customer service" should be translated into goals like, "Improve on-time shipments from 65 percent to 95 percent within two years," or "Reduce lead time from two weeks to four hours on final assembly orders within the next 12 months." Each objective should be measurable and have a target time for completion.

Also critical in securing management backing is the process of preparing a formal justification demonstrating that the financial gain derived from those benefits will justify the cost. We've never found any company that had any difficulty in estimating what the costs are; the real trick is identifying the financial gain that will come about from the new system. It is vitally important to consolidate this formal analysis into three or four main points that will be easily grasped by everybody in the organization, so that everyone will be willing to take steps to make the concept into a working, personal reality.

The next step in this backing is the allocation of sufficient resources to be able to implement the MRP system. This is a difficult task, because the people that are the most critical to our success are the ones that can least be spared to work on something new. We need to find some way of redistributing to underutilized areas the work done by these key employees. This might even involve the hiring of temporary employees to handle the daily work while the permanent employees edit the data going into the new system. Also critical here is making the time needed to conduct the education and training of all the people who are going to be involved with this new system.

We are then ready to create a project team, which should be staffed with people who can be effective in getting a job done. This team is tasked with establishing a project plan that includes specific milestone dates and activities needed to achieve our objectives. Management should actively monitor the progress of the project team and may need to reallocate resources as conditions change over the course of the project.

MANAGEMENT INVOLVEMENT

The second major element of management commitment begins at this point: ongoing management involvement. The purchase of the software system and its installation on the computer is not a guarantee of success. The hardest thing for most managers to understand is that we are moving away from an informal system to a formal method where our actions are managed "by the numbers." This is done most effectively when management participates in the formal planning process. This means the development of a strategic business plan that can be translated into a production plan. The production plan must be quantified in such a way that sales by product line can be matched with our productive capacity. In most discrete manufacturing companies, this also involves going into the next level of master schedule rates and rough-cut capacity plans. While management may not be involved in the mathematics of these individual calculations, there should be active participation in the meetings among marketing, manufacturing, and finance personnel.

The process doesn't end with scheduling a monthly meeting, however. Rather, it is an ongoing coaching process, where management sets responsibility lines—who is supposed to do what—and establishes accountability. Accountability means the process of communicating to the responsible people that they are the clear owners of the process, and setting up lines of communication ensuring that the accountable people can ask for the tools they need to get the job done. This whole procedure is best accomplished if we have a recognizable set of benchmarks telling us where we are now and establishing performance measures to clearly demonstrate our progressive improvement.

THE USER

This brings us to the second element of implementation success, which is user commitment. This is where the real key to success lies. Users must accept ultimate responsibility for the success of the system. The people who make the day-to-day decisions must perceive that this new system is the tool they must have in order to perform their job duties successfully. We need to remember that people make decisions about what actions need to be taken; the computer merely assimilates and reports the effects of those decisions. A computer is only a tool, a mathematical crunch device with a printer attached. People are still, and always will be, the real key in any business.

How do we, as users, get this job done? One simple suggestion is a user checklist like the one in Figure 14–1. The first element is education. As users, we need to take an active role in the techniques involved in using a manufacturing planning and control system. We need to know the techniques for scheduling and maintaining the priorities of materials. In addition, we need to understand the importance of capacity planning and effective control of the shop throughput.

The second element is participation in the development and implementation of the system. As our education progresses, we should be comparing what we do now to the way the new system does it. In some cases, the new software does the job better than our current method. At those points, we should implement the new system as it stands, and let it improve our operations. In some cases, however, our company methods may be better than the generic software, in which case we should first search for ways to work around the system shortcomings. The very last resort is reprogramming. We has-

FIGURE 14–1
User Checklist

1. Education
   - Manufacturing planning scheduling
   - Capacity planning and control
2. Participate in system
   - Development
   - Implementation
3. Operating procedures
4. Training
5. Instill disciplines
6. Ongoing
   - Education
   - Improvements
   - Training

ten to add that most of today's closed-loop MRP systems are very thorough in their base logic and allow workarounds to be managed through printing special reports from a common database. We should give it a lot of thought before interfering with the complicated interrelationships of the base logic of the system by reprogramming for our unique requirements.

The user must, third, be involved in the development of operating procedures. Operating procedures are not three-ring binders that read like legal textbooks and fill shelves in managers' offices; procedures should be written in clear, concise language, covering the basic responsibilities of *who* should do *what* and *when*.

After the procedures are written, we, as users, must train ourselves to use those procedures to solve our problems. We should also spend time learning how to communicate with this new computer system, and learn how to make the system give us the answers to our questions.

Only when we have done this can we move to the next step, where we need to instill the disciplines that will help us achieve data integrity and accuracy. As you will infer from all that we have said to this point, the accuracy of the data is essential if the system is to accurately reflect our needs.

One last point should be made about our user checklist. Participation must be ongoing. This includes a continuing education in new manufacturing techniques and the development of improvements in our existing system. We should never neglect the process of retraining so that we can implement any new improvements. After all, we expect to be in a mode of continuous improvement.

EDUCATION

We have been talking quite a lot about education, and now we need to examine the subject in some depth. Education operates in three areas. The first is the planning and control of the entire business function. Here is where we look at new techniques that are being developed and used at other companies and at how we may be able to apply them to our own situation. The second level of education is what we call "system specific." How does our computer system catalog the information that enables us to achieve our objectives? The third area involves specific training. What particular actions do what particular things? Here, we translate our operating procedures into data management actions.

The education needs to take place at five separate levels of involvement. First comes top management. Management levels are as varied as companies themselves. For our purposes, top management is on a corporate level that may not even be located in the same geographical area as the manufacturing facility. These people need an education focused differently from the one that will be received by middle management. For our use, middle management is resident at the facility and is in charge of managing the day-to-day operations. Both of these levels need to deal with the overall objectives of the system, the allocation of resources, and the expectations of the MRP system's benefits to the enterprise. We have seen education plans that devote anywhere from 4 to 40 hours to these issues. Many companies use outside educators with programs that average around three days and communicate management's involvement in and backing of the system.

The next level is the supervisory force. The supervisors are directly involved in the hour-by-hour operation of the business. They need to know how to manage the people and the flow of data in a way that will enable them to achieve the objectives for which they are responsible. We see most educational programs for this segment range from 30 to 60 hours, allowing these people to become personally familiar, in a hands-on manner, with the system.

The fourth level is what we call the professional level. We're using the concept *professional* in a sense not usually associated with the term; to us, it has to do with attitude and a sense of pride in your work. These are the technically competent people in the organization—the folks who are pivotal in the company's information flow. A description sometimes used for them is *knowledge brokers*. These are the people who absolutely have to understand how to use the new system to solve business problems, and they have to really believe that this tool is their way to get the job done. Here, training must be intensively hands-on and problem-resolution oriented and can take from 40 to 90 hours—or more than that, if necessary.

The fifth level of educational effort encompasses the rest of the work force, which can be seen as two population segments: data entry staff and direct labor. Data entry people need to know how to perform their particular tasks. Everybody else needs some degree of education to demonstrate how their actions fit into the overall company information flow. This can take anywhere from 2 to 10 hours of time, depending on the software and its complexity.

CONVERTING TO THE NEW SYSTEM

From education, we move smoothly (at least on paper) into the actual conversion process. There are a number of ways that can be used to convert from an old system to a new system. The easiest one to remember is "cold turkey." This is where we operate on the old system on one day, and the next morning, we come in and start up operations on the new system. This is considered by most authorities to be catastrophic to both the company and the career of the advocate of this method. Very few successful implementations use this methodology.

An alternate method is running parallel systems; that is, keep the old system running, bring in the new system at the same time, and perform duplicate data entry and produce duplicate reports. One significant drawback to this method is time. If our people are filling eight-hour shifts running the old system, they aren't going to have much time to spare getting a new one running up to operational speed. Besides, there is a subtlety involved in running parallel: If we get two different suggestions from old and new systems, which do we follow? Which system will people trust? People tend to stick with the old, familiar ways of doing things. And if the suggestions are the same, then have we improved our situation by installing a new system?

A much better way is the pilot approach. There are two different pilot techniques. One, known as the *boardroom pilot,* suggests that we install the computer system essentially as we intend to implement it, but with, perhaps, a smaller, "test" database. Several terminals are set up in a training room so that all the users can process transactions as they would in their actual departmental functions. We think that the company's own specific products and situations should be used as users learn to interact with the system to solve their problems. As people progress through the pilot test, we can select a particular product line to become our *real-life pilot.* It is an excellent idea to implement the new system on one complete product line first. It should be a broad enough line so that we can process the product through the entire facility and interact with all types of operations. We should try to move as quickly as we can through the pilot program and extend the new system into the entire operation.

A fourth way of converting is to progressively implement one software module at a time. Practically, this might mean that we are further along in entering bills of material before we become involved in implementing the capacity requirements planning reporting functions. We need to be very cautious here about interfaces and be sure that we understand how new program modules link together and whether we can implement them while the old system is still functioning in other areas. The secret here is to implement the separate modules progressively and quickly to arrive at a fully integrated closed-loop system.

While many different sources deal with implementation, its precise methods, and timing, we generally feel that quicker is better than longer, and once the process is begun, it's better to get on with it. Some milestones we would consider appropriate in judging our performance are, for example, that our pilot program should be going pretty strong after about 9 months from the commitment point, and we should be able to see some real, measurable results in about 12 months. In 18 months to two years, we should have a complete and fully functioning closed-loop system.

PROJECT TEAM

The essential ingredient in a successful conversion is the project team. When we discussed the importance of top-management backing, we mentioned the allocation of the resources and the selection of the project team; it is their efforts, from that point forward, that drive the conversion. There are three elements of this team that we want to examine, as shown in Figure 14–2.

In most companies, we find a great deal of success comes from creating what we might call a steering council. It is composed of top and middle management—the people who have the most to gain in the implementation's success. It usually includes the top management at this geographical location, along with key staff members from other functions. A member of this council should be the project team leader. In most companies, this is a key *user* and somebody we probably can't afford to release from current responsibilities. But this team leadership needs to be a full-time position, whose first priority is the successful implementation of the system. If we are going to achieve the timetable we discussed earlier, two years of half time equals four years on the calendar, and a two-year project that takes four years to implement is unlikely to yield any measurable improvements. This is where we need to bite the bullet and assign a key person to accomplish the job in the shortest time possible. Several experienced sources believe that this person should report directly to top management.

FIGURE 14–2
Project Team

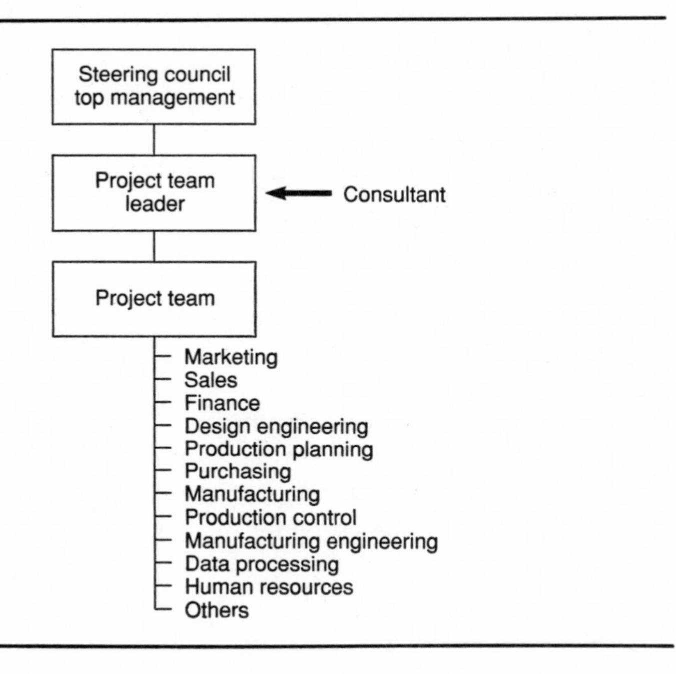

The project team must be made up of key users. These should be people who are capable of making decisions on behalf of their departments; they should have the competence to understand how the new system can help their departments implement their responsibilities, and they should be able to bring to the team a clear description of the functions the system must have in order to serve their departmental needs. Of course, these people also need to be very familiar with our business. These users should participate in the project development and in the pilot so that they can communicate from both departmental and team perspectives. Problems whose resolution exceed team members' authority may have to be referred to the steering council.

The project team members may come and go as various steps of the project are defined and implemented, and their time commitments may vary from department to department over the life of the project. It is important that all functional areas having a stake in the system's success be participating members of the project team. These areas are shown in Figure 14–2, and will be discussed in detail in the next chapter.

Some people advocate delegating responsibility for the project implementation to an outside consultant. The advantages to this approach are the technical expertise, experience, and detachment that a consultant can introduce into the implementation. There is, however, a powerful disadvantage in that perceived ownership of the system resides outside our own company. In addition, there is real value added to our people as they progress through the learning curve.

To get some of the advantages of outside help, it might be useful to solicit the consultant's assistance in helping the project team gain experience. The consultant should work closely with the team leader in those areas where outside expertise is most needed.

Many companies run into trouble by defining a project team and letting it go on forever, remaining constantly in the implementation mode. The key to success is not to get hung up in the implementation process but to focus on the major business benefits that will be accomplished as a result of implementing the system, as emphasized in Focus 14–A. We think it is important to pick out three or four of these major objectives and convert them into measurable goals. Remember that a goal is a measurement and an associated time. We can set performance

FOCUS 14–A
Focus for Success

Focus on major business benefits that are achieved by implementing the system

FIGURE 14–3
Success Strategies

- Focus on major objectives
- Provide measurable goals
- Use performance measures
- Communicate
- Everyone's actions move measurements toward goals

measurements to show us where we are and where we are going; we should probably do this in each functional area. These expectations must then be communicated to everyone throughout the organization. Some specific performance measures are discussed in Chapter 15. We should expect that everyone's actions will move us closer to those goals. An example of this success strategy is shown in Figure 14–3.

APPLICATIONS

In world-class organizations, people tend to internalize these plans and measurable goals to an astonishing degree. We know of a retail and wholesale distributor of wallpaper whose inventory manager got involved in a capacity planning and control education course. Internalizing the performance-measurement concept, he set up a way to measure his stockroom throughput and set up capacity planning, input-output reports, and backlog queues. His objective was to reduce the lead time it took to pick and ship orders. He was able to effectively balance his work schedules to increase the stockroom output. This is a good example of the use of material and capacity planning techniques in a retail/distribution environment. The real winner was, of course, the customer.

CHAPTER 15

INTEGRATING RESOURCES

We have been talking about all the technical issues involved in an MRP system, but the real success comes from the way in which real people—you and I—use these tools. Before we discuss all of the different groups of people in our organizational chart, we need to make sure that we all understand what the goal is. We are striving to become a company exemplifying world-class excellence. It is important that everybody in the organization understand that philosophy. It can be described as being superior in three areas, as shown in Figure 15–1. We must have superior quality and quick delivery at the lowest cost. Let us examine each of these to show how they relate to what we have been saying about an MRP system.

QUALITY

Superior quality is demonstrated in two ways: the quality of our product and the quality of our information. There is a wealth of information from many sources dealing with product quality. We don't need to review it here except to point out that quality should imply that we meet or exceed our customers' requirements. In MRP, we have stressed quality of information. It is imperative that we have a high level of accuracy in our data. The way most people improve their information quality is by printing various reports sorted by specific fields so that knowledgeable people can review the information to determine the validity of the data.

FIGURE 15–1
World Class Excellence

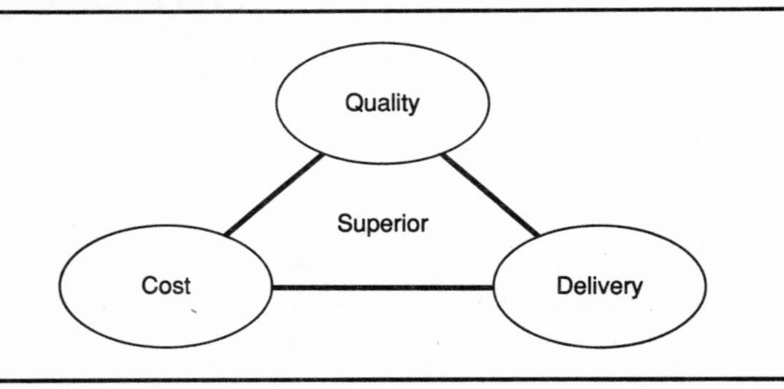

As an example, if we were to print our item master in order by planned lead time, those items showing a zero planned lead time should mean that an order could be started and finished in less than one day. Any parts with very long lead times should be investigated to see if the data is incorrect, or if there are things we can do to reduce that lead time.

In the same process, we should see whether certain groups of parts that should have the same lead times do, in fact, reflect accurate data. The main point of this is to emphasize to our people that our data integrity level should be very high, and that this high level of integrity depends on the individual worker to keep it accurate.

The technique of sorting by lead time and taking corrective actions is one of a series of what we may consider "health indicators" of how well we are managing our MRP system. Several other numbers we should be watching are the levels in our bills of material, the number of items on exception reports, and the number of reschedule notices produced. Two of our favorites are the number of orders past due—purchase orders and manufacturing orders. Management should attempt to keep all of these numbers as low as possible.

These health indicators should not be confused with performance measures, which we will discuss later in this chapter.

There are three critical health indicators that are frequently used as performance measures as well, because they have to be extremely accurate if our MRP system is to function effectively. These elements are shown in Figure 15–2. It is absolutely essential to our MRP success that we measure the accuracy of our inventory records, our bills of material, and our ability to execute the master production schedule. There have been volumes written on how to measure inventory accuracy, the most common method being cycle counting. This is the method where we compare our actual physical count to the book balance on hand. The percentage of instances where the actual and the book counts match should be at least 95 percent in order for us to expect any degree of reliability from our MRP record. Every MRP planning worksheet starts out with the current balance on hand from which we compute projected balances. If we start out with a bad number, we're bound to get bad results.

Let us emphasize that 95 percent is only a place that we're passing through on the journey to higher standards. Cycle counting should be a tool that we have in place to help us isolate and correct the causes of problems in our processes. We should be continuously improving those processes. In our opinion, the standard for inventory accuracy should be about the same as the human standard for a healthy body—someplace in the area of 98.6 percent.

The second area of data integrity is bill of material accuracy. This is where we compare the computerized bill of material to

FIGURE 15–2
Keys to Successful MRP

Data Integrity	*Minimum Accuracy Level*	*Continuous Improvement*
Inventory records	95%	98.6%
Bills of material	98%	100%
Master production schedule	85%	95%

the component list that is actually used in production. Some companies have many different bills of material and can argue for hours about which is right. In our opinion, the "correct" BOM is the way in which the shop puts the product together. This is because those are the parts that are really being consumed from the inventory, which means that those are the parts that need to be planned to come into inventory to fill our next order. If there is a difference in the parts that our people use to build the product and the way that engineering designed the product, one or the other has to be changed so that we can get the two to match up. The gauge of bill of material accuracy is that there should be at least a 98 percent match between the way the product is put together and the bill as it exists in the computer. We should have a process in place to change the computer bill to absolutely match the way the shop assembles the product, since we need to move toward 100 percent accuracy here. Most of our friends in engineering tell us that it's not possible to get 100 percent bill of material accuracy. That's all right. We don't want all of the bills right, just the ones for things that we make.

The third area of data integrity is the master production schedule. Here, we should be striving toward a 95 percent schedule adherence, by which we mean that the quantity that was planned to be produced on a certain day was, in fact, produced on that day. As we discussed in Chapter 6, the MPS is best "firmed up" when it crosses the planning time fence.

DELIVERY

Turning to the second element of our plan for excellence, let's examine quick delivery. The biggest weakness of an MRP system is when people begin to believe that the planned lead times are fixed—that we must adhere to the level-by-level processing that the MRP goes through to build each component and put it into stock before issuing it to the next higher level. As we have said in several other places, we should focus our attention on reducing lead time; that is, reducing the cycle time within all of our processes. This speed applies not only to the production of the product but to the flow of information as well. When a customer calls in to inquire about the delivery of an assembly, we should be able to refer to our MRP system to do what-if analyses or look up component availability to ascertain quickly if that customer's needs can be met.

We think that today's concept of the "internal customer" exemplifies both these concepts of quality and speed for products as well as information. As parts flow from one work center to the next in our shop, it is important that each operator thinks about passing output to the downstream work center's operator as though that next operator were a *customer*. This same principle applies to the flow of information from person to person. We all put data into the MRP system, and so we are supplying that data to our customers, the other system users.

COST

The third element in the excellence journey is cost reduction. We are all very aware of programs that will reduce the costs of our products. Volumes have been written about the emphasis on reducing the direct labor content. Direct labor is a very small portion of the cost of a given product; it is the cost of the materials that usually represents the greater portion and overhead is a large part as well (although we don't know of anyone using the MRP system to reduce that yet). What we should use MRP for is to reduce the material costs through more economical order patterns, which typically reduce the inventory and therefore the carrying costs. There is one other subtle area dealing with costs and the impact of the MRP system on costs that has to do with the people dealing with the flow of parts. In some companies, this is called *expediting* or *coordinating*. One approach is that people are expediting and coordinating individual customer orders to achieve the quickest delivery. While this works for a while in smaller companies, as the numbers of customers and products increase, the process of expediting becomes increasingly complex. One solution is to hire more people to personally handle each job. This, of course, adds to the overhead costs. A far better approach is to have a fully integrated MRP system that allows everybody in the company to work from the same set of information, allowing them to function as a team, shipping product to satisfy the most customers.

This can be summarized simply, as in Focus 15–A: We should all focus on superior quality, quick delivery, and low cost. Each and every action that we take within our MRP system should be directed toward achieving these three goals. Not only must we keep this in mind, but all of the people throughout our organization should be working toward these three ends as a team.

We will now examine each of the functional departments of our organization to see how they interface with our MRP system. We will discuss the benefits each group can expect to derive from the system and what it must contribute to the system to forward this synergy. We will discuss the groups in the order in which they become involved in the product life cycle. Remember that each functional department is made up of our most important resource—people.

FOCUS 15–A
Focus for Success

Focus on:

- Superior quality
- Quick delivery
- Low cost

MARKETING

Marketing gets considerable benefits from a formal MRP system. At one common place, usually at the production plan level, it can compare the expected sales forecast to the manufacturing level of production. If marketing is reasonable in the balance of supply and demand, it can expect a more reliable flow of material and components to support that plan. The price it must pay for this, of course, is that it must have a reasonable way to define and maintain a forecast. We discussed these concepts in detail in Chapter 10 when we examined two-level master scheduling.

SALES

The biggest benefit sales can expect is an improvement in on-time shipping performance to the customer. A corollary of this is the improvement of the accuracy and reliability of our promises to our customers. This is the objective of the MRP system: supplying components to support the master production schedule. In return, sales needs to work "within the system," using such techniques as configured bills and available-to-promise information to help satisfy customer requirements. It should look for ways to help the customer, *not* by creating a unique thing for a single customer, but by trying to use standard products to fill the customer's needs. Sales also has the obligation of establishing good communication lines with our customers, to enable us to avoid surprises and replanning.

FINANCE

The biggest benefit that finance will see from the MRP system is the ability to convert material schedules into financial plans. As we have discussed, the forecast can be multiplied by the unit prices to determine sales revenues. Scheduled receipts plus planned order releases for all purchased items can be used to define the accounts payable requirement. Our capacity loads compared to our staffing levels can be used to anticipate our payroll distribution. In return, we may have to completely rethink the ways in which we accumulate costs. We should look at the MRP system as an opportunity to look at costs at the macro level rather than on a transaction-by-transaction basis.

DESIGN ENGINEERING

Design engineering is where the product is actually designed. If we are using a *CAD/CAM* (computer-aided design/computer-aided manufacturing) system, we can obtain component part information from our MRP system. We can bring prototypes to market much faster by using current components when possible. This should help the design engineer to accelerate the design and introduction of new products. Of course, this means that the design engineer must manage and maintain the bill of material within the computer system, which may require special training to acquire the skills needed to add, modify, or delete the product structure. In addition, the design engineer has to think farther into the future, so that we can design products that are simple to plan and build. This is known as *design for manufacturability*.

PRODUCTION PLANNING

Production planning is where we see one of the biggest benefits of an MRP system, in that we are able to take the production plan and the master production schedule and synchronize the two. We do this with one database, allowing us to be sure that we have planned to build enough product to supply the marketing forecast. The real benefit comes in being able to examine hard data, as opposed to emotional and conflicting opinions, to ensure that we adhere to schedule. The price we pay for this is that we must manage the parameters that show us what the resources are that we need to implement this perfect plan. We need to convert the plan into capacity requirements, financial requirements, and, using lead times, material requirements. Then we need to do whatever it takes to implement that plan. Some people call this "running the business by the numbers," which is contrary to the current culture of many companies.

PURCHASING

The main benefit that purchasing receives from the MRP system is the view into the future that the system gives of the requirement of parts procured from any supplier. As we use the MRP system to maintain valid priorities, purchasing can expect a more accurate set of requirements that the suppliers must fulfill. Another major benefit that purchasing will begin to see from an MRP system is that the users of the system will begin to accept the mindset that the supplier is not an adversary, but rather an extension of our own manufacturing capacity. At the same time, it is important for purchasing to begin utilizing blanket purchase orders, to work with suppliers to schedule capacity instead of scheduling individual and specific products. This is probably the key to success in negotiating more flexibility from the suppliers and working toward a shorter purchasing lead time. We see many companies striving to reduce the complex communications channels that exist between their purchasing and production departments and the suppliers' sales and production departments. This may include the revolutionary idea that our production planners should be dealing directly with the suppliers' production planners. This role usually has the title buyer/planner. The buyer/planner can handle the day-to-day scheduling needs, which allows our purchasing people to spend more time on negotiation and also to focus on a program of supplier certification. A certified supplier has demonstrated the ability to consistently meet quality, cost, and delivery objectives.

MANUFACTURING

Manufacturing has a simple and specific benefit from an MRP system—namely, having the materials when it needs them. Moreover, with capacity planning, it gets a preview of the work that is coming so that it can maintain an efficient flow through the manufacturing process. Manufacturing can also plan and schedule time for preventive maintenance programs. The price it must pay is that it must now plan to be very flexible in its capacities. Not only must we be able to move people to where the work is, we must be able to do it very quickly in order to maintain a very quick throughput rate.

PRODUCTION CONTROL

Production control typically must manage the day-to-day flow of parts and work orders through the manufacturing process. Its biggest gain comes from being able to view the work ahead of time, in the confidence that the priorities are being maintained by the system, so that it is working on the right part at the right time. In order to do this, it is absolutely essential that production control tell the system the entire truth, all the time, every time. If there is a problem, we need to find who can solve it and get us back on schedule.

MANUFACTURING ENGINEERING

Manufacturing engineering normally looks at our processes for actually producing the product. Our MRP database can provide it with information not only about which parts are run on which work centers but also about "where used" capability and the ability to sort items and look at standard costs. The price for this, of course, is the responsibility for maintaining the routings. In addition to the data manipulation, manufacturing engineering must also work on shortening the lead time through our various manufacturing processes.

DATA PROCESSING

Data processing usually looks at the MRP system with great rejoicing. It is a significant advantage to have an integrated system where the various modules work together so that data entered by users in one functional area is accessible to users from other areas. Most systems also provide report-writer capabilities to allow users to create their own ad hoc reports. There is a real dilemma in all of this. Users are entering data and writing their own reports and accessing on-line, real-time data, which can take a heavy toll on our data processing efficiency, not to mention data security. Data processing will have to find ways of migrating from the efficiency of batch processing to allow users access to instantly updated information. To do this, data processing professionals will need to move from report-writing skills into the areas of security, maintenance, and hardware efficiency.

HUMAN RESOURCES

The biggest benefit of the MRP system to human resources is that it gives education and training programs a rallying point around an integrated system. With this integrated system, we can focus more readily on the company objectives of providing customer service. The price to pay, most often, is that there is going to be a change in the company culture as we move toward a more flexible work force with broader skill bases, with a view toward a more empowered employee.

TEAMWORK

In order for all of these people to work together, we need to set our minds on teamwork, and how we can develop attitudes that will facilitate teamwork. The best way to describe this is that management should take a coaching approach. How can we reconcile the idea that management should run the business by the numbers, in the formal way, while still focusing on coaching and encouraging the worker? There are three ways in which this can be accomplished, as shown in Figure 15–3. These means are functional accountability, MRP effectiveness, and performance measures.

Functional accountability means that all of the groups of people we discussed above should be given goals and objectives that are consistent with one another and that accomplish the company's overall objectives. We need to communicate these objectives throughout the organization, which we do by training our people to utilize the system to make decisions. We also need to educate the people as to what decisions they should make and how their actions improve the operation of the company. Third, we must motivate the people to strive toward continuous improvement in company performance.

How do we use the MRP system to do these things? Let's examine our MRP effectiveness checklist in Figure 15–4. There are three main areas. First, plan and control inventory. This means that our system should order the right part at the right time and in the right quantity. We should take steps occasionally to check out a few items at random to make sure that the

FIGURE 15–3
Management as Coach

1. Functional accountability
2. MRP effectiveness
3. Performance measures

FIGURE 15–4
MRP System Effectiveness Checklist

1. Plan and control inventory
   - Order the right part
   - At the right time
   - In the right quantity
2. Maintain priorities
   - Order with proper due date
   - Maintain due date that matches need date
3. Input to capacity
   - Future visibility
   - Accurate dates on load
   - Total load

system really is doing its job in this regard. We shoud also check to see that the safety stocks are maintained only at the master production schedule level.

The second element of our checklist is that the MRP system needs to maintain priorities. Our first concern should be to see that every part is ordered with the proper due date. Second, the due date should be flexible—as conditions change, the system must make sure that the due date of the order matches the need date of the requirement. We must also remember to keep the master schedule realistic. The integrity of the system depends on not overloading the MPS. If we can maintain that integrity, we can expect that the MRP will be valid and describe all the actions we need to take to support the master schedule.

The third item on our effectiveness checklist is to use our MRP schedules as an input to the capacity plan. The objective is to have clear visibility of what our workload is expected to be on future dates. We must, therefore, have accurate dates on all the orders. This load would then be the complete load, representing scheduled receipts plus planned order releases.

In the real world, managers of effective MRP systems take the time to periodically check some data to make sure that these three elements are in place.

Another way managers determine whether we are on the right track is performance measurement. Here, we are entering an area where many companies need to develop measures appropriate to their own products and processes. However, there are some common areas that we should make sure we have covered, as shown in Figure 15–5.

We have already discussed the first two items on our list, inventory record and bill of material accuracy. Looking at the subject of schedule adherence, we should examine the master schedule items to see how many were produced on the day they were supposed to be produced. This concept should be extended to work orders and supplier deliveries as well.

The next measurement is the one that is probably most important to our customers—on-time shipping performance. In our opinion, this should be the percentage of orders that were shipped on the date that the customer requested. To compare against our shipping performance, we should examine our inventory level, which includes finished goods, work in process, and raw material inventories. Comparing this to our cost of goods sold, we can compute inventory turns.

Another area we feel is extremely important in managing

FIGURE 15–5
Performance Measures

Measure	*Target*
Inventory record accuracy	98.6% of records
Bill of material accuracy	100%
Schedule adherence	
MPS orders—end units	95% quantity on date
Work orders—components	95% complete on date
Supplier deliveries	95% on time
On-time shipping performance	99% on time by customer request date
Inventory level	Turns
Lead time	100% at/below standards

an effective MRP system is lead time. We should have a way of measuring the percentage of orders completed in our shop on or before the lead time standard.

Of course, there are other measures not shown in Figure 15–5 that are not directly related to our MRP system—for example, quality of our products and various operating costs.

The purpose of all these performance measurements is not to have people competing against one another to make the numbers look good. Rather, these numbers should be consistent with each other; our ultimate objective is to have all of our people working together to effect continuous improvement in our products and processes. Management should focus its attention on the people to mold together a world-class team, as shown in Focus 15–B. Today's companies must create a vision of excellence to demonstrate how they are continually improving quality, delivery time, and cost reduction, as shown in Figure 15–6.

No matter where we are in our journey to world-class excellence, the race is never finished. As we see in Figure 15–7, we must strive constantly to climb higher and get better. An MRP system is not the total answer; it is one foundation block that teaches us how to run our business in an organized, formal, and logical manner. Once we can do this, then we are able to go on to total quality control (TQC), just-in-time, and computer-integrated manufacturing (CIM) systems. The integration of MRP with our total business enterprise is a major first step in the migration toward more advanced technology.

FOCUS 15–B
Focus for Success

Focus on the *people* to mold together a *world-class team.*

FIGURE 15–6
World-Class Excellence

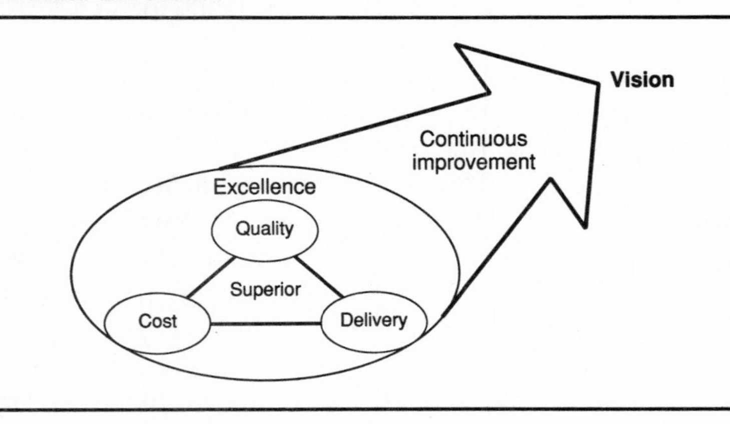

APPLICATIONS

How do we pull all this together, get the vision before all our people, and set performance measures that are consistent? One of the best examples we know of is a service industry that has effectively utilized all of these concepts even though it doesn't actually manufacture a product. The airfreight forwarding com-

FIGURE 15–7
Technology Migration Steps

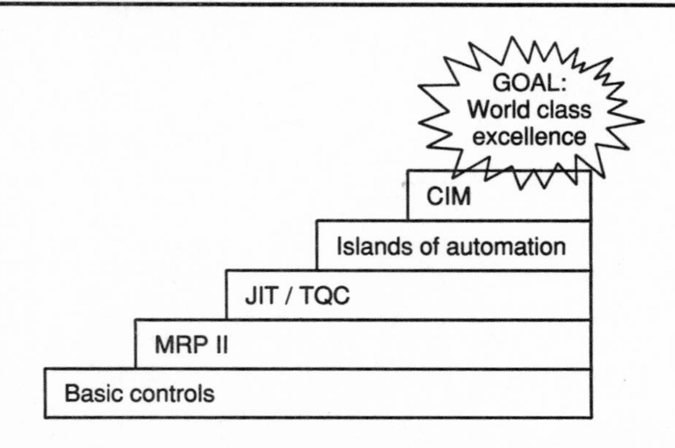

pany that we have mentioned several times demonstrates through measurement its consistent near-perfect record of on-time deliveries to its customers. Its record-keeping system enables it to pinpoint the location of a single package at any time during its transit.

One of the ways the airfreight company does this is by posting a scoreboard in every area of the company. This scoreboard posts the percentage of packages that reached customers on time and the percentage of airplanes that left the ground on schedule. If the airfreight company can do this while tracking in excess of a million packages daily, any manufacturing plant should be able to measure its on-time shipments and schedule adherence.

We also know a manufacturing company that measures its sales rates by piece and then uses this information to release orders. Out on the manufacturing floor, it keeps track of the number of parts released and the number of parts produced each day. Everybody in the organization can quickly identify the expected rates for each production line. The company measures quantities of parts produced and the average lead time that it took. That lead time, again, is posted on the manufacturing floor for everyone to see. While the company did more than install an MRP system, in less than two years it has successfully shortened lead time from more than three months to less than two weeks.

Both of these companies have attained exceptionally high levels of customer service by motivating their people to use formal systems to manage the business.

Notice that both these companies have visual systems that communicate clear goals that the entire work force team can rally around and identify with. The people in the company have a clear vision of their current position on the road toward continuous improvement.

And where is your company? Can you apply any of the things we have discussed to move farther along that road?

We wish you a productive journey.

APPENDIXES

APPENDIX A

SUGGESTED READING

Since the primary purpose of this book is to give the reader a solid foundation in the basic logic of MRP and to establish the position of material and capacity requirements planning within the business hierarchy, there is a vast amount of related information that we have touched on only briefly or omitted from this book altogether. Readers who need to know more about these areas, or who have specific applications problems outside the scope of this book, might find some of the following reading suggestions helpful. This is not an all-inclusive list, and many fine texts may have been omitted from it; what we hope to do here is point you in the direction of some of the many valuable resources provided by a few representative members of our industry.

For more acclimatization in the broad area of production and inventory management, we suggest:

Fogarty, Donald W.; John H. Blackstone, Jr.; and Thomas R. Hoffmann. *Production and Inventory Management.* 2nd ed. Cincinnati: South-Western Publishing, 1991.

Plossl, George W. *Production and Inventory Management: Principles and Techniques.* 2nd ed. New York: Prentice-Hall, 1985.

Vollmann, Thomas E.; William L. Berry; and D. Clay Whybark. *Manufacturing Planning and Control Systems.* 3rd ed. Homewood, Ill.: Business One Irwin, 1992.

For assistance in the selection and implementation of an MRP II system, the Oliver Wight Companies have a series of informative publications. To get started, you might find these titles helpful:

Gray, Christopher D. *The Right Choice: A Complete Guide to Evaluating, Selecting, and Installing MRP II Software.* Essex Junction, Vt.: Oliver Wight Limited Publications, 1987.

Wallace, Thomas F. *MRP II: Making It Happen—The Implementers' Guide to Success with Manufacturing Resource Planning.* 2nd ed. Essex Junction, Vt.: Oliver Wight Limited Publications, 1990.

Wight, Oliver W. *The Executive's Guide to Successful MRP II.* Essex Junction, Vt.: Oliver Wight Limited Publications, 1985.

For more detailed material on capacity planning, two informative sources are:

Blackstone, John H., Jr. *Capacity Management.* Cincinnati: South-Western Publishing, 1989.
Correll, James G., and Norris W. Edson. *Gaining Control: Capacity Management and Scheduling.* Essex Junction, Vt.: Oliver Wight Limited Publications, 1990.

Bills of material are examined in detail in two books:

Garwood, Dave. *Bills of Material: Structured for Excellence.* Marietta, Ga.: Dogwood Publishing, 1988.
Mather, Hal. *Bills of Material.* Homewood, Ill.: Business One Irwin, 1987.

The past several years have seen the publication of literally hundreds of books on just-in-time (JIT), ranging from discussions of JIT philosophy to implementation guidebooks to detailed treatments of particular applications of JIT in many areas—purchasing, maintenance, accounting, setup and cycle time reduction, employee empowerment, and plant layout, to name just a few. Some of the many books that have assisted companies we know about to move toward a world-class way of doing business include:

Claunch, Jerry W., and Philip D. Stang. *Set-Up Reduction: Saving Dollars with Common Sense.* Palm Beach Gardens, Fla.: PT Publishing, 1989.
Hall, Robert W. *Attaining Manufacturing Excellence.* Homewood, Ill.: Business One Irwin, 1987.
Wantuck, Kenneth A. *Just in Time for America.* Southfield, Mich.: KWA Media, 1989.

Another area that is finally receiving long-overdue attention from many gifted contributors is quality and customer service. The reader unfamiliar with the varying philosophies and techniques involved in establishing and evaluating quality programs would profit from reading one or more of the "quality gurus" or their disciples. To test the waters, we suggest:

Crosby, Philip. *Quality Is Free.* New York: McGraw-Hill, 1979.
Deming, W. Edwards. *Out of the Crisis.* Cambridge, Mass.: MIT Press, 1982.
Juran, Joseph M. *Juran on Planning for Quality.* New York: Free Press, 1988.

Walton, Mary. *The Deming Management Method.* New York: G. P. Putnam's Sons, 1986.

Once an MRP system has been selected and installed, it should not be treated as a free-standing mechanical device churning out masses of paper in some remote office. We have suggested that there are ways in which the MRP system can become a valued tool to enhance our management process. Some interesting insights on this and on preselection of software, with helpful advice to implementation teams, can be found in:

Luber, Alan D. *Solving Business Problems with MRP II.* Bedford, Mass.: Digital Press, 1991.

An excellent book covering many areas that businesses should be attuned to as they move toward world-class performance is:

Moody, Patricia E., ed. *Strategic Manufacturing: Dynamic New Directions for the 1990s.* Homewood, Ill.: Business One Irwin, 1990.

If we are to remain competitive and creative, we should never stop learning as much as we can about our jobs and the various resources available to us. We commend the certification programs administered by the American Production and Inventory Control Society (APICS), together with the many books, articles, and courses available through APICS, on the national and local levels, that support those programs. For membership, publication, or certification information, write:

American Production and Inventory Control Society, Inc.
500 West Annandale Road
Falls Church, Virginia 22046

APPENDIX B

PEN–AND–PENCIL MRP EXERCISE

The purpose of this exercise is to give you the opportunity to practice doing an MRP by hand. Of course, we must start with the three inputs to our MRP system. The first is the master production schedule. Figure B–1 is the master schedule for Pens 42-208. Figure B–2 shows Pencil 43-208. You will notice that we do not have the forecast or the full information to create the master schedules, so you can use the basic master schedule without any complexities. In the real world, it won't be this simple, but here, you should be concentrating your full attention on the mechanics of the MRP.

The second input is the bills of material. Figure B–3 is the bill of material tree for the pen, and Figure B–4 shows the pencil BOM. To help you understand the data on these trees, for each part we show the short name for the part, and underneath that, the part number. Underneath the part number we note the "quantity per," or the amount needed to build one assembly. You can assume the quantity per to be 1 when nothing is noted. We have also provided the indented explosions for the pen in Figure B–5 and for the pencil in Figure B–6, if you wish to use them.

The third main input to our MRP is the inventory status. Figure B–7 shows the FineLine Industries inventory status sorted by part number, giving us the lead time, lot size, and current on-hand balance for each of the parts that concerns us. Figure B–8 is FineLine's inventory status for open orders, again for those parts we are interested in. You will notice that it shows the quantity of the part number that is on order and the period in which it is scheduled for delivery.

PART A

Run the MRP. We need to prepare an MRP worksheet (Figure B–9 is a double blank for photocopying) for each part that is required to sup-

FIGURE B–1
Master Production Schedule

Pen 42-208 — Period

Gold, blue ink	Now	1	2	3	4	5	6	7	8
Forecast									
Actual orders									
On hand	0								
Available-to-promise									
Master schedule		20			20		10		25

FIGURE B–2
Master Production Schedule

Pencil 43-208 — Period

Gold, thin lead	Now	1	2	3	4	5	6	7	8
Forecast									
Actual orders									
On hand	0								
Available-to-promise									
Master schedule			10		10	30			25

FIGURE B–3
Bill of Material for Pen 42-208

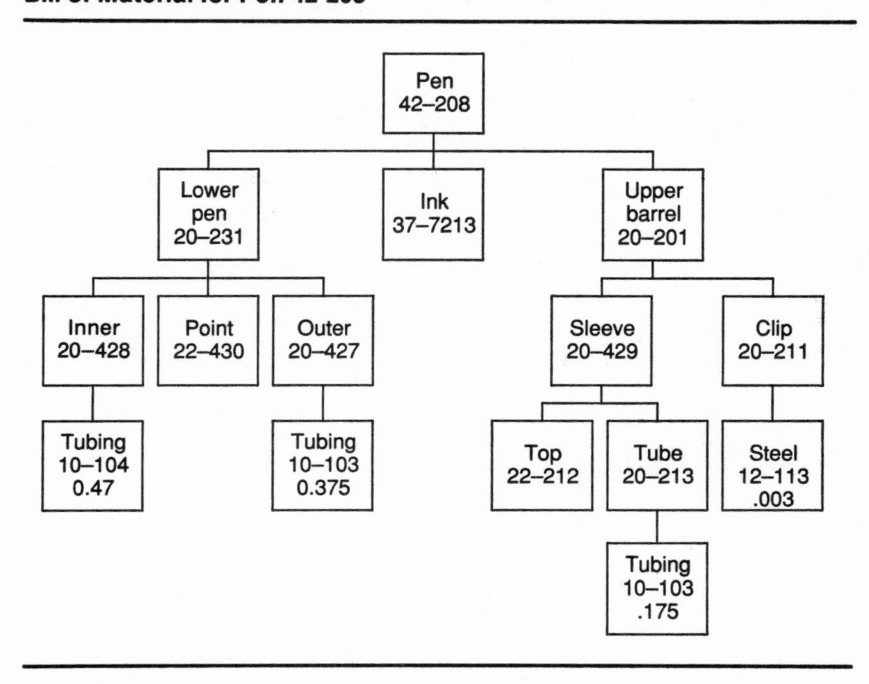

port the master schedule for these pens and pencils. Some assumptions that you need to make are:

1. Our master schedules represent the total requirement.
2. We are creating a regenerative MRP.
3. Work to have minimum inventory while using the lot sizes that are given for each part.
4. To simplify the mathematics, assume that all scrap factors are zero. You can work to the nearest 10th of a unit of measure.
5. *All* open shop orders have allocations that have *not* been filled.

As you are preparing these MRP worksheets, be sure to note what reschedule notices would appear on an exception report. Also be sure to note what new orders should be released now.

FIGURE B–4
Bill of Material for Pencil 43-208

Level 0 / 1 / 2 / 3

- Pencil 43-208 (Level 0)
  - Upper barrel 20-201 (Level 1)
    - Sleeve 20-429 (Level 2)
      - Top 22-212 (Level 3)
      - Tube 20-213 (Level 3)
        - Tubing 10-103 176
    - Clip 20-211 (Level 2)
      - Steel 12-112 .003 (Level 3)
  - Lower pencil 20-241 (Level 1)
    - Inner tube 20-235 (Level 2)
      - Tubing 10-104 .489 (Level 3)
    - Outer tube 20-236 (Level 2)
      - Tubing 10-103 .375 (Level 3)
    - Lead 16-108 (Level 2)
      - Long leads 16-100 .066 (Level 3)
    - Point 22-431 (Level 2)
  - Eraser 18-108 (Level 1)
    - Eraser 18-109 (Level 2)
      - Material 18-101 .525 (Level 3)
    - Socket 18-110 (Level 2)
      - Steel 12-108 .001 (Level 3)
  - Lead 16-108 4 (Level 1)
    - Long leads 16-100 .066 (Level 2)

FIGURE B–5
Indented BOM Explosion (Pen)

Parent 42-208: Pen, gold, blue ink; 1 each

Level	*Component*	*Quantity Per*	*Description*
1	20-231	1.000 each	Lower subassembly, gold pen
.2	20-428	1.000 each	Inner sleeve, gold pen
..3	10-104	0.470 feet	Tubing, inner
.2	20-427	1.000 each	Outer tube, gold pen
..3	10-103	0.375 feet	Tubing, outer casing
.2	22-430	1.000 each	Point, gold pen, medium
1	20-201	1.000 each	Upper barrel, gold
.2	20-429	1.000 each	Outer sleeve, gold
..3	20-213	1.000 each	Upper barrel tube
...4	10-103	0.175 feet	Tubing, outer casing
..3	22-212	1.000 each	Upper barrel top
.2	20-211	1.000 each	Upper barrel clip
..3	12-113	0.003 pound	Steel, spring clips
1	37-7213	1.000 each	Ink cartridge, blue

FIGURE B–6
Indented BOM Explosion (pencil)

Parent 43-208: Pencil, gold, thin lead; 1 each

Level	*Component*	*Quantity Per*	*Description*
1	20-241	1.000 each	Lower subassembly, gold pencil
.2	20-235	1.000 each	Inner sleeve, gold pencil
..3	10-104	0.489 feet	Tubing, inner
.2	20-236	1.000 each	Outer tube, gold pencil
..3	10-103	0.375 feet	Tubing, outer casing
.2	22-431	1.000 each	Point, gold pencil, thin
.2	16-108	1.000 each	Lead, thin, 3/4″ long
..3	16-100	0.066 feet	Long thin lead
1	20-201	1.000 each	Upper barrel, gold
.2	20-429	1.000 each	Outer sleeve, gold
..3	20-213	1.000 each	Upper barrel tube
...4	10-103	0.175 feet	Tubing, outer casing
..3	22-212	1.000 each	Upper barrel top
.2	20-211	1.000 each	Upper barrel clip
..3	12-113	0.003 pound	Steel, spring clips
1	18-108	1.000 each	Eraser assembly
.2	18-109	1.000 each	Eraser 1/2″ long
..3	18-101	0.525 inch	Eraser material 1/4″ diameter
.2	18-110	1.000 each	Eraser socket
..3	12-108	0.001 pound	Steel eraser socket
1	16-108	4.000 each	Lead, thin, 3/4″ long
.2	16-100	0.264 feet	Long thin lead

FIGURE B–7

FineLine Industries Inventory Status (part-number order)

Part Number	*Pen and pencil items only* Description	Lead Time	Lot Size	Current On-Hand Balance
10-103	Tubing, outer casing	4	10	25
10-104	Tubing, inner	4	50	15
12-108	Steel, eraser socket	5	800	0
12-113	Steel, spring clips	6	10,000	0
16-100	Lead, long thin	5	6 feet	0
16-108	Lead, thin, 3/4″ long	1	1	70
18-101	Eraser material 1/4″ diameter	0	7 feet	0
18-108	Eraser assembly	Phantom		0
18-109	Eraser 1/2″	Phantom		0
18-110	Eraser socket	2	300	180
19-002	Instructions book			
19-086	Box cover, gold pen set			
19-087	Box base			
19-088	Box platform, gold			
19-089	Gold name plate			
19-090	Gold price sticker			
20-085	Box, gold pen set			
20-201	Upper barrel, gold	2	25	50
20-210	Upper barrel, sleeve, gold			
20-211	Upper barrel clip	2	600	207
20-213	Upper barrel tube	4	100	25
20-231	Lower subassembly, gold pen	2	25	50
20-235	Inner sleeve, gold pencil	4	30	0
20-236	Outer tube, gold pencil	Phantom		0
20-241	Lower subassembly, gold pencil	1	30	20
20-427	Outer sleeve, gold pen	Phantom		0
20-428	Inner sleeve, gold pen	5	200	2
20-429	Outer sleeve, gold	Phantom		0
22-212	Upper barrel top	5	25	25
22-430	Point, gold pen, medium	5	50	0
22-431	Point, gold pencil, thin	3	50	0
37-7213	Ink cartridge, blue, medium point	7	50	50
37-7215	Ink cartridge, black			
42-208	Pen, gold, blue ink	MPS		0
42-211	Pen, gold, black ink			
42-377	Pens, planning bill			
43-208	Pencil, gold, thin lead	MPS		0
51-101	Set, gold pen and pencil			

FIGURE B–8
FineLine Industries Inventory Status (open orders)

Open orders — Pen and pencil items only

Quantity	*Description*	*Part Number*	*Scheduled Delivery*
10 pieces	Outer tubing	10-103	3
110 feet	Inner tubing	10-104	1
6 feet	Long thin lead	16-100	3
6 feet	Long thin lead	16-100	4
178 feet	Eraser material This is a blanket order with remaining balance unscheduled	18-101	
25 pieces	Upper barrel	20-201	5
25 pieces	Upper barrel tube	20-213	2
30 pieces	Inner pencil sleeve	20-235	3
25 pieces	Upper barrel top	22-212	2
50 pieces	Point, gold pen	22-430	5

FIGURE B–9
MRP Worksheet

	Now	Period 1	2	3	4	5	6	7	8	9
Requirement										
Scheduled receipts										
On hand										
Planned order releases										

	Now	Period 1	2	3	4	5	6	7	8	9
Requirement										
Scheduled receipts										
On hand										
Planned order releases										

FIGURE B–10
FineLine Industries Exception Reports

Reschedule Notices

Description	*Part No.*	*Quantity*		*Period*
Point	22-430	50	from	5 to 6
Upper	20-201	25	from	5 to 4
Top	22-212	25	from	2 to 3
Tube	20-213	25	from	2 to 3
Inner pencil	20-235	30	from	3 to 4
Tube	10-103	10	from	3 to 8

6 items total of 9 possible = 67%

Planned Order Releases To Release Now

Description	*Part No.*	*Quantity*		*Period*
Ink	37-7213	50	now	Due 8
Top	22-212	50	now	Due 6
Pencil point	22-431	50	now	Due 4
Long leads	16-100	6 feet	now	Due 6
Inner tube	20-428	200	now	Due 6

5 items total of 21 possible = 24%

Note: Before you move on to the problems that follow in Part B, you should check the solution for this MRP run. The exception report is given in Figure B–10, and the filled-in worksheets are Figures B–11 through B–31.

FIGURE B–11
MRP Worksheet

20-231
Lower Pen
LT=2 Lot=25

	Now	1	2	3	4	5	6	7	8	9
Requirement		20			20		10		25	
Scheduled receipts										
On hand	50	30	30	30	10	10	0	0	0	0
Planned order releases							25			

FIGURE B–12
MRP Worksheet

22-430
Pen point
LT=5 Lot=50

	Now	1	2	3	4	5	6	7	8	9
Requirement							25			
Scheduled receipts						50				
On hand	0	0	0	0	0	50	25	25	25	25
Planned order releases										

Suggest: Reschedule 50 due 5 Move out to 6

FIGURE B–13
MRP Worksheet

20-428
Pen inner
LT=5 Lot=200

	Now	1	2	3	4	5	6	7	8	9
		Period								
Requirement							25			
Scheduled receipts										
On hand	2	2	2	2	2	2	177	177	177	177
Planned order releases		200								

Suggest: Release order 200 now due 6

FIGURE B–14
MRP Worksheet

10-104
Tubing
LT=4 Lot=50

	Now	1	2	3	4	5	6	7	8	9
		Period								
Requirement	14.7	94		14.7						
Scheduled receipts		110								
On hand	15	16.3	16.3	1.6	1.6	1.6	1.6	1.6	1.6	1.6
Planned order releases										

FIGURE B–15
MRP Worksheet

20-427
Outer pen
Phantom

	Now	Period 1	2	3	4	5	6	7	8	9
Requirement							25			
Scheduled receipts										
On hand	0	0	0	0	0	0	0	0	0	0
Planned order releases							25			

FIGURE B–16
MRP Worksheet

10-103
Tubing, outer
LT=4 Lot=10

	Now	Period 1	2	3	4	5	6	7	8	9
Requirement	4.4		17.5		11.3		9.4	11.3		
Scheduled receipts				10						
On hand	25	20.6	3.1	13.1	1.8	1.8	2.4	1.1	1.1	1.1
Planned order releases			10	10						

FIGURE B–17
MRP Worksheet

37-7213
Ink cartridge
LT=7 Lot=50

	Now	1	2	3	4	5	6	7	8	9
Requirement		20			20		10		25	
Scheduled receipts										
On hand	50	30	30	30	10	10	0	0	25	25
Planned order releases		50								

Suggest: Release 50 now due in 8

FIGURE B–18
MRP Worksheet

20-201
Upper barrel
LT=2 Lot=25

	Now	1	2	3	4	5	6	7	8	9
Requirement		20	10		30	30	10		50	
Scheduled receipts						25				
On hand	50	30	20	20	-10	10	0	0	0	0
Planned order releases				25			50			

Suggest: Reschedule 25 due in 5 Move in to 4

FIGURE B–19
MRP Worksheet

20-429
Upper sleeve
Phantom

	Now	1	2	3	4	5	6	7	8	9
Requirement	25			25			50			
Scheduled receipts										
On hand										
Planned order releases	25			25			50			

FIGURE B–20
MRP Worksheet

20-212
Upper barrel top
LT=5 Lot=25

	Now	1	2	3	4	5	6	7	8	9
Requirement	25			25			50			
Scheduled receipts			25							
On hand	25	0	25	0	0	0	0	0	0	0
Planned order releases		50								

Suggest: Reschedule 25 due 2 Move out to 3
Order 50 now due in 6

FIGURE B–21
MRP Worksheet

20-213
Upper tube
LT=4 Lot=100

	Now	1	2	3	4	5	6	7	8	9
Requirement	25			25			50			
Scheduled receipts			25							
On hand	25	0	25	0	0	0	50	50	50	50
Planned order releases			100							

Suggest: Reschedule 25 due in 2 Move out to 3

FIGURE B–22
MRP Worksheet

20-211
Clip
LT=2 Lot=600

	Now	1	2	3	4	5	6	7	8	9
Requirement	25			25			50			
Scheduled receipts										
On hand	207	182	182	157	157	157	107	107	107	107
Planned order releases										

FIGURE B–23
MRP Worksheet

18-109
Eraser 1/2"
Phantom

	Now	1	2	3	4	5	6	7	8	9
Requirement			10		10	30			25	
Scheduled receipts										
On hand										
Planned order releases			10		10	30			25	

FIGURE B–24
MRP Worksheet

18-101
Eraser material
LT=0 Lot=7

	Now	1	2	3	4	5	6	7	8	9
Requirement			5.3		5.3	15.8			13.2	
Scheduled receipts										
On hand	0	0	1.7	1.7	3.4	1.6	1.6	1.6	2.4	2.4
Planned order releases			7		7	14			14	

Remarks: 178 Feet on blanket order

FIGURE B–25
MRP Worksheet

20-241
Lower pencil
LT=1 Lot=30

	Now	1	2	3	4	5	6	7	8	9
Requirement			10		10	30			25	
Scheduled receipts										
On hand	20	20	10	10	0	0	0	0	5	5
Planned order releases					30			30		

FIGURE B–26
MRP Worksheet

22-431
Pencil point
LT=3 Lot=50

	Now	1	2	3	4	5	6	7	8	9
Requirement					30			30		
Scheduled receipts										
On hand	0	0	0	0	20	20	20	40	40	40
Planned order releases		50			50					

Suggest: Order 50 now due in 4

FIGURE B–27
MRP Worksheet

20-235
Inner pencil tube
LT= 4 Lot= 30

	Now	1	2	3	4	5	6	7	8	9
Requirement					30			30		
Scheduled receipts				30						
On hand	0	0	0	30	0	0	0	0	0	0
Planned order releases				30						

FIGURE B–28
MRP Worksheet

20-236
Outer pencil tube
Phantom

	Now	1	2	3	4	5	6	7	8	9
Requirement					30			30		
Scheduled receipts										
On hand										
Planned order releases					30			30		

FIGURE B–29
MRP Worksheet

16-108
Leads 3/4" long
LT=1 Lot=1

	Now	1	2	3	4	5	6	7	8	9
		Period								
Requirement			40		40 30	120		30	100	
Scheduled receipts										
On hand	70	70	30	30	0	0	0	0	0	0
Planned order releases				40	120		30	100		

FIGURE B–30
MRP Worksheet

16-100
Long thin lead
LT=5 Lot=6

	Now	1	2	3	4	5	6	7	8	9
		Period								
Requirement				2.6	7.9		2.0	6.6		
Scheduled receipts				6	6					
On hand	0	0	0	3.4	1.5	1.5	5.5	4.9		
Planned order releases		6	6							

Suggest: Order 6 now due in 6

FIGURE B–31
MRP Worksheet

18-110
Eraser socket
LT=2 Lot=300

	Now	1	2	3	4	5	6	7	8	9
		Period								
Requirement			10		10	30			25	
Scheduled receipts										
On hand	180	180	170	170	160	130	130	130	105	105
Planned order releases										

PART B

Here are two problems that you need to address in your role as the planner for pens and pencils at FineLine Industries:

1. The supplier of the inner tubing, Part 10-104, says that its order for 110 feet, due this week, shipped short. FineLine will only receive 100 feet. The supplier can put the other 10 feet into its next production run, but that's not until four periods from now. What actions would you recommend to the supplier? What other actions should we take?
2. The supplier of long thin lead, Part 16-100, says that the open order for 6 feet scheduled to ship in Period 3 will be delayed until Period 4.

Rerun the MRP with the suggestions that you feel we should use to solve these two problems.

Solutions to these problems are found on p-ages 280–288.

SOLUTIONS TO PROBLEMS

Problem 1

In Figure B–33, we are looking at the inner tubing, Part 10-104, where the vendor has short-shipped 100 feet instead of 110 feet. We would use pegging to see the source of the requirement in Period 1, which would lead us to the inner sleeve, Part 20-428, as shown in Figure B–32. You can see that the requirement was because of the large lot size. The solution is to create a firm planned order for 175 pieces instead of 200, thereby reducing inventory. We can cancel the remaining 10 feet of tubing on order. *Note:* We should reduce this order so that we can keep some raw material available for the requirement in Period 3.

The lesson we have learned from this is that it was our lot-size inventory that caused the large requirement for tubing. We should always look at the intermediate levels as we try to solve problems with an MRP system.

Problem 2

There are two options for solving this problem when our supplier of pencil leads is unable to deliver: Option A is to use full pegging to go to the master schedule level and move the 10 pencils planned for Period 4 out to Period 5. We would only do this if we have explored all the intermediate levels. If we make that change to the master schedule and rerun the MRP, we can see the results in Figures B–34 through B–42. As you can see, there aren't as many changes as we might first think because of the effect of lot sizing. However, there is one serious negative effect to our customer service. If we had all of the information on the master schedule, we would know whether it was unacceptable to our customers, in which case we could solve our problem using Option B. This solution is shown in Figures B–43 and B–44. Remembering that MRP lead times are planned lead times, we can often compress the manufacturing lead time. If we look at the leads, Part 16-108, we can see that the lead time is one period. We need to talk with manufacturing and set a firm planned order for 40 pieces in Period 4, with a lead time of zero.

FIGURE B–32
MRP Worksheet

20-428
Pen inner
LT= 5 Lot=200

	Now	1	2	3	4	5	6	7	8	9
Requirement							25			
Scheduled receipts										
On hand	2	2	2	2	2	2	152	152	152	152
Planned order releases		175								

Firm up order for 175 due in 6

Pegging

FIGURE B–33
MRP Worksheet

10-104
Tubing
LT= 4 Lot= 50

	Now	1	2	3	4	5	6	7	8	9
Requirement	14.7	82.3 ~~94~~		14.7						
Scheduled receipts		100 ~~110~~								
On hand	15	18.4	18.4	3.7	3.7	3.7	3.7	3.7	3.7	3.7
Planned order releases										

FIGURE B–34
Master Production Schedule

Pencil 43-208 Gold, thin lead	Now	Period 1	2	3	4	5	6	7	8
Forecast									
Actual orders									
On hand	0								
Available-to-promise									
Master schedule			10		(10) →	30			25

FIGURE B–35
MRP Worksheet

20-241 Lower pencil LT=1 Lot=30	Now	Period 1	2	3	4	5	6	7	8	9
Requirement			10		(10) →	30			25	
Scheduled receipts										
On hand	20	20	10	10	10 ~~0~~	0	0	0	5	5
Planned order releases					30			30		

FIGURE B–36
MRP Worksheet

20-201
Upper Barrel
LT=2 Lot=25

	Now	1	2	3	4	5	6	7	8	9
Requirement		20	10		~~30~~ 20	~~30~~ 40	10		50	
Scheduled receipts						25				
On hand	50	30	20	20	~~10~~ 0	10	0	0	0	0
Planned order releases				25			50			

Suggest: ~~Reschedule 25 due 5 Move in to 4~~
Do Not Reschedule

FIGURE B–37
MRP Worksheet

16-108
Leads 3/4" long
LT=1 Lot=1

	Now	1	2	3	4	5	6	7	8	9
Requirement			40		(40) 30	120		30	100	
Scheduled receipts										
On hand	70	70	30	30	0	0	0	0	0	0
Planned order releases				(40)	120		30	100		

FIGURE B–38
MRP Worksheet

16-100
Long thin lead
LT=5 Lot=6

		Period								
	Now	1	2	3	4	5	6	7	8	9
Requirement				2.6	7.9		2.0	6.6		
Scheduled receipts				6	6					
On hand	0	0	0	~~3.4~~ 0	1.5	1.5	5.5	4.9		
Planned order releases		6	6							

Suggest: Order 6 now due in 6

FIGURE B–39
MRP Worksheet

18-109
Eraser 1/2"
Phantom

		Period								
	Now	1	2	3	4	5	6	7	8	9
Requirement			10		10	30			25	
Scheduled receipts										
On hand										
Planned order releases			10		10	30			25	

FIGURE B–40
MRP Worksheet

18-101
Eraser material
LT=0 Lot=7

	Now	1	2	3	4	5	6	7	8	9
Requirement			5.3		(5.3) →	15.8			13.2	
Scheduled receipts										
On hand	0	0	1.7	1.7	~~3.4~~ 1.7	1.6	1.6	1.6	2.4	2.4
Planned order releases			7		~~7~~	~~14~~ 21			14	

Remarks: 178 feet on blanket order

FIGURE B–41
MRP Worksheet

18-108
Eraser assembly
Phantom

	Now	1	2	3	4	5	6	7	8	9
Requirement			10		(10) →	30			25	
Scheduled receipts										
On hand										
Planned order releases			10		(10) →	30			25	

FIGURE B–42
MRP Worksheet

18-110
Eraser socket
LT=2 Lot=300

	Now	1	2	3	4	5	6	7	8	9
Requirement			10		(10) →	30			25	
Scheduled receipts										
On hand	180	180	170	170	~~160~~ 170	130	130	130	105	105
Planned order releases										

FIGURE B–43
MRP Worksheet

16-108
Leads 3/4" long
LT=1 Lot=1

	Now	1	2	3	4	5	6	7	8	9
Requirement			40		40 30	120		30	100	
Scheduled receipts										
On hand	70	70	30	30	0	0	0	0	0	0
Planned order releases				~~40~~	~~120~~ 160		30	100		

Firm planned order with LT=0 40 pieces due in 4

FIGURE B–44
MRP Worksheet

16-100
Long thin lead
LT=5 Lot=6

	Now	1	2	3	4	5	6	7	8	9
						Period				
Requirement				~~2.6~~	10.5 ~~7.9~~		2.0	6.6		
Scheduled receipts				(6) →	6					
On hand	0	0	0	0 ~~3.4~~	1.5	1.5	5.5	4.9		
Planned order releases		6	6							

Suggest: Order 6 now due in 6

Other Observations

As you worked your way through this exercise, one thing should have become very clear: The cumulative lead time for the components is such that our master schedule horizon does not extend far enough into the future. We must either have master schedule orders further out in the future or we need to shorten some of our lead times. The cumulative lead time for the pen is 11 periods and the cumulative lead time for the pencil is 10 periods, which indicates the minimum length of the planning horizon for our master schedules.

While working through this exercise, it just might "feel" more comfortable to have one big order for each part that would cover the full horizon. This is a common occurrence among planners, but as you can see from some of these items, it's a dangerous way to operate because it builds inventory. It is really much better to try to establish some flow rates and standard delivery cycles. An example of this is the eraser material, where our supplier should see our MRP plan for all those future releases against our blanket purchase order.

There are some items, such as the steel, Part 12-113, that are not listed in our solution. This is because there are no requirements or actions that need to be taken. Actually, it's not necessary to print any of the items that have no suggested actions.

A couple of technical details are worth noting. The planning for the requirements for leads, Part 16-108, demonstrates the principle of low-level coding. Upper barrels, Part 20-201, had an order that was released far too early, resulting in reschedule notices. One other interesting note is that the lead, Part 16-108, has a lot size equal to 1, which implies the lot-for-lot technique.

It is important to note that building lots in excess of requirements creates inventory at multiple levels. This should be avoided when we focus on setup reduction. Also, notice the effect of the phantoms, which compress both the lead time and lot-size inventory.

If you have any interesting solutions or other points, please let us know.

APPENDIX C

LOT–SIZING TECHNIQUES

For the reader who is interested in mathematical detail, here is some expanded information on the nine most commonly used lot-sizing techniques. In order to help us to visualize the effects of these lot-sizing techniques, we are going to use one set of data, as shown in Figure C–1.

Notice that the periods represent months, which we chose in order to ease the calculation of carrying costs per period. Of course, with a computer, we can be precise to the *n*th decimal point and use weekly periods.

You should also note that the requirement line represents net requirements. We will compute the lot size and record the projected on-hand balances as though the balance on hand always starts at zero. Note also that the lead time for manufacture is zero in all cases, which Means that the material arrives in the same period in which we order it. We are assuming that setup costs us $120 per order, each unit costs $20, and our company has a policy of 24 percent annual carrying costs.

FIXED LOT SIZE

In fixed lot sizes, we set one number that is constant, which is usually derived from some restriction in the method of supply. In our example, shown in Figure C–2, the fixed lot size is 100 pieces and the resulting two orders in nine months leave remnants, or leftovers, in inventory in each period. This lot size has the advantage of being the easiest to understand but should only be used when the methods of supply are severely restricted.

FIGURE C–1
MRP Worksheet

Lead time = 0
Setup = $120
Unit cost = $20

		Month								
	Now	1	2	3	4	5	6	7	8	9
Requirement		5	40		10		50	25	10	40
Scheduled receipts										
On hand	0	−5	−45	−45	−55	−55	−105	−130	−140	−180
Planned order releases										

LOT–FOR–LOT

In this method, we arbitrarily decide to produce a shop order for each period in which there is a net requirement, and the quantity of that shop order will exactly match the net requirement. In Figure C–3, you will see that we have seven orders in our nine-month period. This gives us minimum inventory investment, and we have zero balance on hand throughout the planning horizon. This becomes the lot size of

FIGURE C–2
MRP Worksheet (fixed lot size)

Lead time = 0
Setup = $120
Unit cost = $20

		Month								
	Now	1	2	3	4	5	6	7	8	9
Requirement		5	40		10		50	25	10	40
Scheduled receipts										
On hand	0	95	55	55	45	45	95	70	60	20
Planned order releases		100					100			

Fixed Lot = 100

FIGURE C–3
MRP Worksheet (lot-for-lot)

Lead time = 0
Setup = $120
Unit cost = $20

	Now	1	2	3	4	5	6	7	8	9
Requirement		5	40		10		50	25	10	40
Scheduled receipts										
On hand	0	0	0	0	0	0	0	0	0	0
Planned order releases		5	40		10		50	25	10	40

Month (columns 1–9)

choice for items of extremely high cost. The disadvantage of this method is that it generates a high proportion of orders; we would want to have a very low setup cost to balance this.

FIXED PERIOD

In order to gain economy of scale with items that have significant setup cost, we may wish to group together the net requirements for several periods at a time. Many companies arbitrarily pick the periods, and in Figure C–4, we have elected to produce an order every three months rather than once a month. You will see that the number of orders on the horizon has decreased from seven to three, giving us an improvement in manufacturing efficiency. The price we pay is an increase in inventory, as we have some balance on hand remaining in several periods.

ECONOMIC ORDER QUANTITY (EOQ)

Over the years, many formulas have been derived in an attempt to arrive at an acccurate balance between the high setup costs involved in running many orders as opposed to the cost of holding inventory balance on hand. Several of our reference texts in Appendix A

FIGURE C–4
MRP Worksheet (fixed periods)

Lead time = 0
Setup = $120
Unit cost = $20

	Now	1	2	3	4	5	6	7	8	9
Requirement		5	40		10		50	25	10	40
Scheduled receipts										
On hand	0	40	0	0	50	50	0	50	40	0
Planned order releases		45			60			75		

Periods = 3

go through the mathematical derivation of minimizing the sum of the setup cost and the holding cost. The formula itself is shown in Figure C–5.

Notice the assumption of annual usage, since the mathematics includes the annual carrying cost of 24 percent. The future usage in our example is 180 pieces in nine months, representing 240 pieces per year on the average. Substituting the values into the formula gives us an EOQ of 110. Planning the two orders of 110 pieces each is demonstrated in Figure C–6.

FIGURE C–5
EOQ Formula (economic order quantity—EOQ)

$$EOQ = \sqrt{(2AS/IC)}$$

Where:

A = Annual usage
S = Setup cost ($120)
I = Annual carry cost (24%)
C = Unit cost ($20)

Since:

Then: Annual usage = 180 pieces/9 months × 12 months

$$EOQ = \sqrt{(2 \times 240 \times 120/.24 \times 20)} = 110$$

FIGURE C–6
MRP Worksheet (EOQ)

Lead time = 0
Setup = $120
Unit cost = $20

	Now	Month 1	2	3	4	5	6	7	8	9
Requirement		5	40		10		50	25	10	40
Scheduled receipts										
On hand	0	105	65	65	55	55	5	90	80	40
Planned order release		110						110		

The advantage here is that our total number of orders has been reduced to 2. We have some remnants left over in inventory, as you can see from the balance of five pieces in Period 6. Notice that the second order for 110 pieces gives us an excess in inventory of 40 pieces at the end of the planning horizon. This is because the EOQ formula assumes a continuum of demand and could, therefore, calculate excess inventory. While it is a simple way of comparing holding costs to setup costs, it is only applicable to parts having smooth, uniform, and continuous demand. From an MRP point of view, this type of demand only applies to independent demand items; we expect lumpy demand for dependent demand items, making EOQ a poor lot-sizing choice.

PERIOD ORDER QUANTITY

In our EOQ example, our first order left us remnants in inventory. In order to minimize the effect of this excess inventory, the period order quantity lot-sizing technique was developed. Here, we divide the EOQ into the estimated annual usage to derive an estimated number of orders per year. In Figure C–7, we compute 2.18 orders per year, or about a six months' interval between orders. Therefore, in our example, the first order, covering a six-month period, will be 105 pieces. The second order would be for 75 pieces, covering the rest of our horizon. This gives us the advantage of the EOQ, in that we reduce the number of replenishment orders while avoiding the disadvantage of large remnants in inventory.

FIGURE C–7
MRP Worksheet (period order quantity)

Lead time = 0
Setup = $120
Unit cost = $20

	Month									
	Now	1	2	3	4	5	6	7	8	9
Requirement		5	40		10		50	25	10	40
Scheduled receipts										
On hand	0	100	60	60	50	50	0	50	40	0
Planned order releases		105						75		

Annual usage = 240 per year
EOQ = 110

Implies 2.18 orders per year
or about a 6 months' interval

LEAST UNIT COST

To avoid the assumption of annualized usage, mathematicians have developed two methods of minimizing the sum of setup and holding costs on a period-by-period basis, as opposed to an annual basis. In the first of these, the least unit cost method, we compute first the holding cost per part, per month. In Figure C–8, we see that our item cost of $20 and the 24 percent carrying cost imply 2 percent per month, making our holding cost $0.40 per part per month. Looking at the

FIGURE C–8
Least Unit Cost Chart

I = Annual carry cost 24% per year
Item cost $20 × 0.02 = $0.40 per part per month

	Quantity	*Setup*	*Holding Cost*		*Total*	*Unit Cost*	
Lot 1	5 pieces	$120	0	$ 0	$120	$ 24 each	
Lot 2	45 pieces	$120	40 × .4 =	$ 16	$136	3.02 each	
Lot 3	55 pieces	$120 +	10 × .4 × 3 =	$ 28	$148	2.69 each	
Lot 4	105 pieces	$120 +	50 × .4 × 5 =	$128	$248	2.36 each	←
Lot 5	130 pieces	$120 +	25 × .4 × 6 =	$188	$308	2.37 each	

FIGURE C–9
MRP Worksheet (least unit cost)

Lead time = 0
Setup = $120
Unit cost = $20

		Month								
	Now	1	2	3	4	5	6	7	8	9
Requirement		5	40		10		50	25	10	40
Scheduled receipts										
On hand	0	100	60	60	50	50	0	50	40	0
Planned order releases		105						75		

possible lot sizes, starting in Period 1, adding together the setup cost plus the holding cost to give us the total cost and dividing the result by the lot size, we get the unit cost. As you can see in our various options in Figure C–8, the unit cost continues to fall through Lot 4. Therefore, since that is the least unit cost, the first order will be 105 pieces. Figure C–9 shows the results of this plan. In period 7, we will need to go through the same calculations; the lot size here would be 75.

The advantage of this method is that it allows us to compare the setup costs to the holding costs on a period-by-period basis. It is applicable to items with high cost. A disadvantage is the amount of computer time required to do the calculations.

LEAST TOTAL COST

To overcome the computer data-crunching problems, mathematicians developed the concept of least total cost. Here, we calculate the ratio of the setup cost to the unit holding cost, as shown in Figure C–10. You can see that the setup cost is divided by the product of the monthly carrying percentage times the unit cost. This is a nondimensional ratio of 300, called the *economic part period*. Similar to the least unit cost method's evaluation of several possible lot sizes, this least total cost method also gives us a list of alternatives. Instead of having to do all the calculations, we only need to track the number of parts times the

FIGURE C–10
Least Total Cost Chart

Economic part periods = Setup/(Carry × Unit cost)
EPP = \$120/(0.02 × \$20) = 300

	Quantity	*Part Periods*		*Accumulated*
Lot 1	5 pieces	0	0	0
Lot 2	45 pieces	40 × 1 =	40	40
Lot 3	55 pieces	10 × 3 =	30	70
Lot 4	105 pieces	50 × 5 =	250	320 ←
Lot 5	130 pieces	25 × 7 =	175	495

total number of periods; this is the element called Part Periods shown in Figure C–10. We accumulate part periods until they total a number greater than the economic part period ratio. In our example, we "break" the 300 economic part periods when the lot size reaches 105, so our first order will be for 105 in Period 1. We would go through the same calculations to derive a second lot of 75 in Period 7. We show the orders and the projected balance on hand in Figure C–11.

This method yields the same result as least unit cost, but the mathematical method used to derive it is slightly different.

FIGURE C–11
MRP Worksheet (least total cost)

Lead time = 0
Setup = \$120
Unit cost = \$20

	Now	Month 1	2	3	4	5	6	7	8	9
Requirement		5	40		10		50	25	10	40
Scheduled receipts										
On hand	0	100	60	60	50	50	0	50	40	0
Planned order releases		105						75		

PART PERIOD BALANCING

This method takes the least total cost method and performs two re-evaluations of that solution. The first is called *look ahead*, the second is called *look back*. Here, we look at the second replenishment order to determine whether it should be moved ahead or back one period in order to change the replenishment order quantity on the first order. This is best demonstrated in the example in Figure C–12, where we perform the look ahead calculation. We look at the next order, which is due in Period 7, and we suppose what would happen if we were to move it ahead to Period 8. If we moved it ahead, we would save the 10 pieces demanded in Period 8 or save them from holding in inventory. But this would cost us, in that we would have to build the 25 pieces needed in Period 7 way back in Period 1, holding them in inventory for six periods. This would cost us 25 pieces for six periods in order to save 10 pieces for one period. Obviously, this is not an economical alternative, and we should not move the order ahead.

Conversely, looking at Figure C–13, we apply the look back method. Here, we look at the order suggested in Period 7, and we move this order back to Period 6. This will cost us the 25 pieces

FIGURE C–12
MRP Worksheet (part period balancing—look ahead)

Lead time = 0
Setup = $120
Unit cost = $20

	Now	1	2	3	4	5	6	7	8	9
		Month								
Requirement		5	40		10		50	25	10	40
Scheduled receipts										
On hand	0	100	60	60	50	50	0	50	40	0
Planned order releases		105						75	xx	

Look ahead: Move order to Period 8
Save 10 pieces 1 period
Cost 25 pieces 6 periods — No

FIGURE C–13
MRP Worksheet (part period balancing—look back)

Lead time = 0
Setup = $120
Unit cost = $20

	Now	Month 1	2	3	4	5	6	7	8	9
Requirement		5	40		10		50	25	10	40
Scheduled receipts										
On hand	0	100	60	60	50	50	0	50	40	0
Planned order releases		105					xx	75		

Look back: Move order to Period 6
Save 50 pieces 5 periods — Yes
Cost 25 pieces 1 period

required in Period 7 to be held in inventory for one period. However, we will save the 50 pieces required in Period 6 that would have been held in inventory for five periods. The savings outweigh the costs, so we agree to the look-back alternative and readjust our planned order back to Period 6. This will also cause the order in Period 1 to be reduced from 105 pieces to 55 pieces.

The primary advantage of part period balancing is that the look-ahead/look-back feature tends to group the orders together at points of large lumps of demand. This method is applicable to high-priced items that enjoy sporadic demand. The disadvantage of the method is its complexity, making it hard to understand; therefore, planners have a tendency to blindly acquiesce to the computer's suggestions.

WAGNER–WHITEN ALGORITHM

Looking at part period balancing, we might object that the look-ahead/look-back feature should evaluate more than one period. This is what the Wagner-Whiten Algorithm is designed to do. It evaluates all the conceivable permutations and combinations of choices to arrive at an optimal order suggestion covering the known demand stream. It is most applicable to products having a limited life, where the demand will come to a dramatic cessation. Fashion items or products with specific obsolescence might use this method to scale down production.

APPENDIX D

CLOSED–LOOP MANUFACTURING SYSTEM

Figure D–1 is the graphic representation of the positions of and interactions between the various functions that comprise a closed-loop manufacturing system. Throughout the book, we have referred to the various pieces of this chart and described some of them in detail. At this point, we have put together, in one place, a discussion of how the pieces fit together. We will also examine some things that are not directly related to an MRP system.

You will notice that there are basically five levels of detail in the top-down planning process. The first level is *long-range planning*, with the strategic business plan being the major item. This is the long-range strategic plan defining the objectives of our company as a whole and can typically cover a range of three to five years or longer. Inputs to this plan include markets, production processes, and financial objectives. More often than not, these plans are financial in nature, using dollars as a unit of measure. This business strategy should be the driving force for all the rest of our production scheduling and capacity management systems. Concurrent with the strategic business plan, we develop our market forecasts. These include several inputs such as extrinsic factors—market conditions external to our company. We are also interested in intrinsic factors, those calculations based on historical averages. To do all of this, we need to evaluate the sources of demand from our various field outlets.

We accumulate these various forecasts and feed the data into a module frequently known as *demand management*. At this point, we also include our *distribution resource planning system*, or *DRP*. This system deals with the replenishment inventory at all of our branch warehouses. We will actually use MRP-type logic to calculate planned replenishment orders. These future plans are also fed into the demand management module. Demand management accumulates the various demands for our products into family summaries. This data is then fed into our *production plan*.

FIGURE D–1
Closed-Loop Manufacturing System

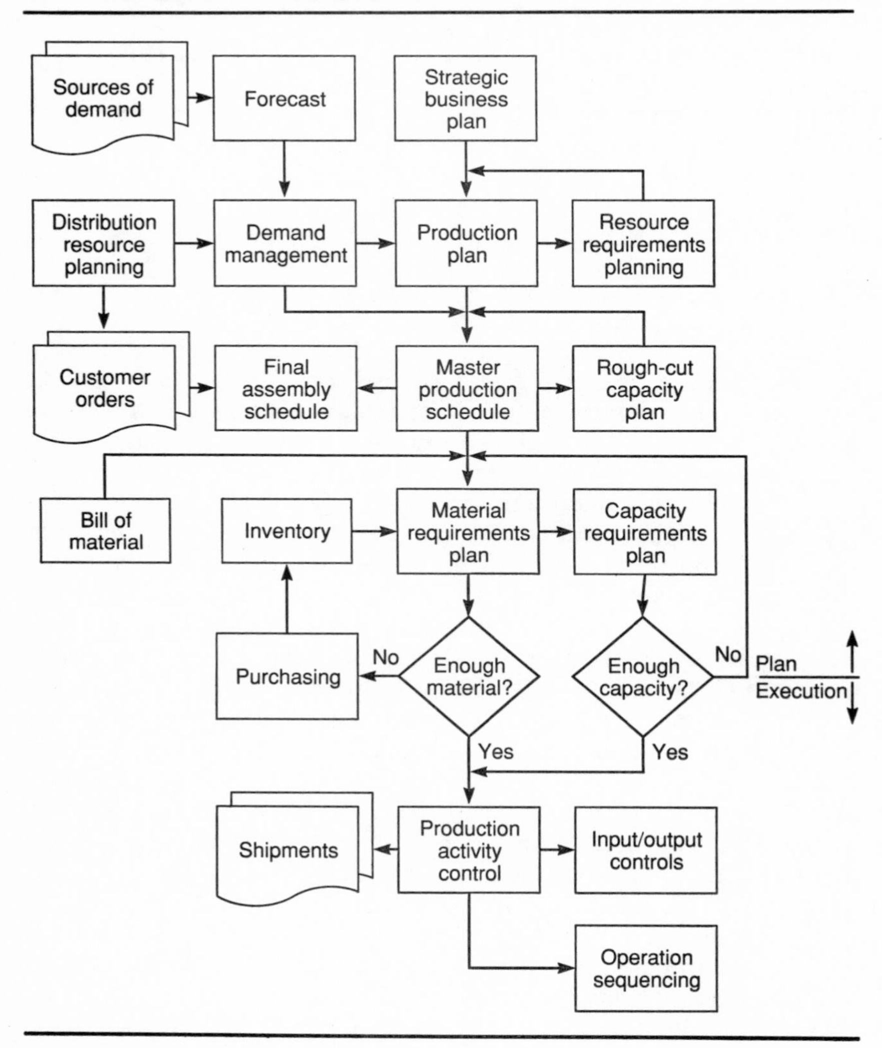

At the production planning level, we accumulate the demand for each family and anticipate the production rates we must set in order to supply that demand. It is important that this aggregate rate of output be consistent with the marketplace demand, and it is also important that the rates of production allow us to achieve our strategic business plan. Here is where we coordinate our sales levels with our production level to achieve the planned inventory level. We know, also, that the family production plans must be verified against our *resource requirements plan*. The resource requirements plan has as its input the historical aggregate data. This data deals with the plants, assembly lines, and manpower that are needed to build our production plan. This is the reason that the arrow is drawn back, so that we check to see that our resources are adequate, using a feedback loop, before we move on to the third planning level.

If the family plans are realistic, then we disaggregate those plans into the *master production schedule*. The master production schedule is for the end units we will produce. The input is the aggregate family plans and actual customer orders, as well as the forecasted demand by individual products. At the same time, the distribution resource planning system and actual customer orders are being fed into the final assembly schedule. The master production schedule should be building parts, components, and end units to feed the final assembly schedule. This master production schedule level of planning should cover a horizon out to about a year or so, depending on the complexity and lead time of our product. It must be long enough to include the cumulative lead time of the components. The output from the master production schedule is the planned orders for specific quantities of specific part numbers to be built on specific dates. This data is fed into a *rough-cut capacity plan*. The purpose of the rough-cut plan is to verify that the master production schedule can be made. Its input is the bills of labor, showing the hours of work required for the master schedule items across critical work centers. This capacity plan will produce an output showing the hours required on the work centers for specific time periods. At this point, it should be reviewed by management to verify that we have enough capacity to build all of the items on the master schedule. If the answer is no, we must either adjust our capacities or rearrange the schedules to achieve the production plan. This is another example of a feedback loop. If, however, we have sufficient capacity, then the master schedule planned orders are passed to the *material requirements planning* level.

At the material requirements planning level, we are dealing with all the component items needed to build the master schedule. Here, we have inputs from the bill of material, defining the parent/component relationships, and from the inventory system, showing the balance on hand and any open replenishment orders. The material requirements plan suggests planned order releases and reschedule notices. All of MRP's planned orders, released orders, and firm planned orders are input into the *capacity requirements plan*. The capacity requirements plan also has the routings as an input. From this data, it can produce a report showing the load by date for every work center. Again, at this level of planning, management should determine if we have enough capacity to be able to build the schedule. If we do not, we must either adjust our capacity, or, as the last resort, rearrange the schedules. If we have insufficient material, we must release orders to purchasing to be able to feed components to inventory. If we have enough purchased materials for the orders and enough capacity, we can then go on to release the orders to our manufacturing department.

We now pass to the fifth and final level, which is execution. For every manufactured item, we release orders to the shop. These orders show specific part numbers, quantities, due dates, and start dates. These orders feed the *production activity control* module, sometimes called *shop floor control*. This is the system that maintains and communicates the status of all the shop orders and work centers. This is where it all comes together. One technique we use to measure the effectiveness of our manufacturing operations is called *input/output controls*. Here, we compare our planned rates of input to our planned rates of output and monitor the resulting queue. We also feed the input/output control module our actual performance data, so we can compare it to our plan and determine whatever corrective action is needed. One tool used to effect this corrective action is called *operations sequencing*. Inputs to this technique include released orders and the staffing levels of our various work centers. We can then use forward finite scheduling to simulate the loads and estimate shop completion dates. Of course, as this all comes together, the finished goods are driven into our shipping room to enable us to fulfill customer demands.

There are other concepts that are universally applicable to this closed-loop diagram. One such concept is quality, and we advocate quality at the source. All employees involved in the activities in Figure D–1 should endeavor to provide perfect quality in each of their processes and outputs.

Another element of utmost importance throughout this diagram is data integrity. With all of this data shifting from module to module, we need to be as accurate as possible. This responsibility doesn't reside just with the data processing department, although it involves them. Data integrity is the responsibility of every person in the organization.

Up and down this diagram, at every level, the performance measures we institute should motivate our entire team toward our unanimous objectives.

Over the years, we have found one summary to be most helpful in understanding and internalizing these principles, as shown here as Figure D–2. The strategic business plan makes up the first level of our diagram, but the objectives of production and inventory control deal with materials and capacities at the four remaining levels. Our objective is to be able to plan the material flow (sometimes called *material priorities* or *material schedules*). At each step through the levels of planning, we need to check the material plan against our capacity. If the capacity is sufficient, we can pass to the next planning level. If it is not, we need to stop where we are and correct the capacity to enable us to

FIGURE D–2
Closed-Loop Diagram

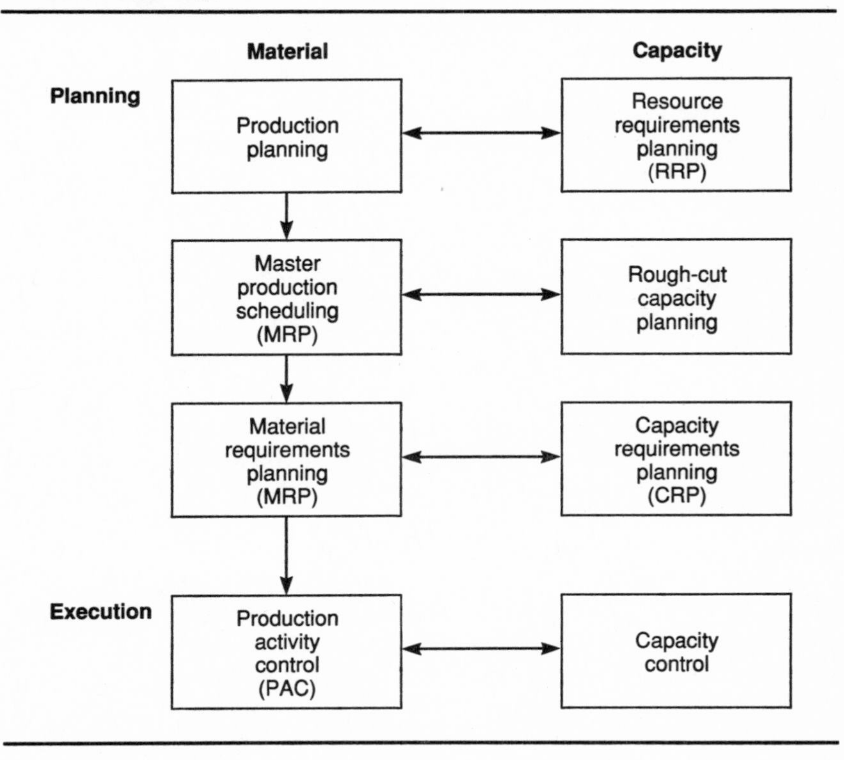

achieve our plan. As shown in Figure D–2, at the production planning, or aggregate family, level, we need to check resource requirements. At the master production schedule, or end unit, level, we check rough-cut capacities. The next level is the individual components plans, as dictated by material requirements planning. Here is where we get the detailed work center capacity requirements plan. If these three levels of planning are correct, we can pass to the execution phase, where production activity control manages the work orders in the shop; the capacity controls will enable our people to effectively manage production.

INDEX

* Definition from *APICS Dictionary, sixth edition,* used by permission of American Production and Inventory Control Society, Inc., Falls Church, Va.

J

K

L

M

N

O

P

About APICS

APICS, the educational society for resource management, offers the resources professionals need to succeed in the manufacturing community. With more than 35 years of experience, 70,000 members, and 260 local chapters, APICS is recognized worldwide for setting the standards for professional education. The society offers a full range of courses, conferences, educational programs, certification processes, and materials developed under the direction of industry experts.

APICS offers everything members need to enhance their careers and increase their professional value. Benefits include:

- Two internationally recognized educational certification processes—Certified in Production and Inventory Management (CPIM) and Certified in Integrated Resource Management (CIRM), which provide immediate recognition in the field and enhance members' work-related knowledge and skills. The CPIM processes focuses on depth of knowledge in the core areas of production and inventory management, while the CIRM process supplies a breadth of knowledge in 13 functional areas of the business enterprise.
- The APICS Educational Materials Catalog—a handy collection of courses, proceedings, reprints, training materials, videos, software, and books written by industry experts . . . many of which, are available to members at substantial discounts.
- *APICS The Performance Advantage*—a monthly magazine that focuses on improving competitiveness, quality, and productivity.
- Specific industry groups (SIGs)—suborganizations that develop educational programs, offer accompanying materials, and provide valuable networking opportunities.
- A multitude of educational workshops, employment referral, insurance, a retirement plan, and more.

To join APICS, or for complete information on the many benefits and services of APICS membership, **call 1-800-444-2742** or **703-237-8344**. Use extension 297.

OTHER BOOKS OF INTEREST TO YOU FROM THE BUSINESS ONE IRWIN/APICS SERIES IN PRODUCTION MANAGEMENT

BENCHMARKING GLOBAL MANUFACTURING
Understanding International Suppliers, Customers, and Competitors
Jeffrey G. Miller, Arnoud De Meyer, and Jinichiro Nakane

Grounded in data collected from survey responses of over 1,000 manufacturing companies worldwide, *Benchmarking Global Manufacturing* reveals valuable insights into these organizations' performance, operations, and strategies. With these comparisons, and with the book's hands-on benchmarking toolkit, you will be equipped to challenge assumptions and think strategically in every decision. (443 pages)

ISBN: 1-55623-674-3

GLOBAL PURCHASING
How to Buy Goods and Services in Foreign Markets
Thomas K. Hickman and William M. Hickman, Jr.

Provides you with the blueprint you need to set up an efficient international sourcing operation—with little or no international purchasing experience. You'll discover ways to reduce labor costs, access cutting-edge technology, and sharpen your competitive edge by sourcing outside of the United States. (237 pages)

ISBN: 1-55623-416-3

COMMON SENSE MANUFACTURING
Becoming a Top Value Competitor
James A. Gardner

Details how you can integrate your manufacturing process and transform your company into a world-class competitor . . . even if you have limited resources. The book's common-sense planning approach leads to quality and service enhancement and well-managed inventory and operations. (249 pages)

ISBN: 1-55623-527-5

MANUFACTURING PLANNING AND CONTROL SYSTEMS
Third Edition
Thomas E. Vollmann, William L. Berry, and D. Clay Whybark

In this Third Edition, state-of-the-art concepts and proven techniques are combined to offer a practical solution to enhancing the manufacturing process. Each of the book's central themes—Master Planning, Material Requirements Planning, Inventory Management, Capacity Management, Production Activity Control, and Just-In-Time—has been updated to reflect the newest ideas and practices. (844 pages)

ISBN: 1-55623-608-5

NO POSTAGE
NECESSARY
IF MAILED
IN THE
UNITED
STATES

BUSINESS REPLY MAIL
FIRST-CLASS MAIL PERMIT NO 2858 FALLS CHURCH, VA

POSTAGE WILL BE PAID BY ADDRESSEE

ATTN: MARKETING DEPARTMENT
AMERICAN PRODUCTION AND INVENTORY
CONTROL SOCIETY INC
500 W ANNANDALE RD
FALLS CHURCH VA 22046-9701